THE AMERICAN FAMILY

Dying or Developing

THE
AMERICAN
FAMILY

Dying or Developing

Edited by
David Reiss, M. D.
George Washington University
Washington, D.C.

and
Howard A. Hoffman, M. D.
The Psychiatric Institute
Washington, D.C.

Plenum Press · New York and London

Library of Congress Cataloging in Publication Data

Main entry under title:

The American family, dying or developing.

Proceedings of the conference held in Washington, D.C. on June 10–11, 1978.
Includes bibliographies and index.
1. Family – United States – Congresses. I. Reiss, David, 1937-   II. Hoffman,
Howard.
HQ536.A547   301.42'0973   78-24447
ISBN 0-306-40117-7

The preparation of this book was supported, in part, by
a grant from the National Endowment for the Humanities.

© 1979 Plenum Press, New York
A Division of Plenum Publishing Corporation
227 West 17th Street, New York, N.Y. 10011

Printed in the United States of America

This book is dedicated to the anonymous donor, whose generous gift supported the writing and editing of this volume. Private donations have long been a main-stay of American scholarship. As Maimonides recognized centuries ago, anonymous donations confer an additional degree of dignity and intellectual freedom on those who receive them.

Contributors

T. Berry Brazelton. Dr. Brazelton is Associate Professor of Pediatrics at Harvard Medical School. He has made significant and influential contributions to the study of the development of infants and mother-infant relationships. His years of searching and affectionate scholarship are epitomized in his recent book *Infants and Mothers: Individual Differences in Development* which received the Child Study Association of America Award.

Martha Cox. Dr. Cox is Assistant Professor of Psychology at the University of Evansville, Evansville, Indiana. She has made significant contributions to the study of the family process, foster families, and marriage.

Roger Cox. Dr. Cox is Assistant Professor of Psychology at Indiana State University. He has made significant contributions to the study of child development. He has authored numerous publications on the study of exceptional children and on child development.

David Brion Davis. Professor Davis is Farnam Professor of History at Yale University. His work on the history of slavery and of European and American reform movements has had a fundamental impact on current concepts and analysis in the field of modern history. He has also made significant contributions to the study of

American intellectual history. His book *The Problem of Slavery in Western Culture* won the 1967 Pulitzer Prize. More recently, another book, *The Problem of Slavery in the Age of Revolution 1770-1823*, was selected by *The New York Times* as one of the 13 best books published in 1975.

BERNICE T. EIDUSON.   Dr. Eiduson is Professor of Medical Psychology in the Department of Psychiatry at the University of California in Los Angeles. She has made significant contributions in the field of child development, varying family styles, and mental health in early development. Dr. Eiduson is the author of several books, among them *Science as a Career Choice*, which she authored with Linda Beckman; and *Scientists: Their Psychological World*. She is on the editorial board of *Developmental Psychobiology*, was formerly on the editorial board of *Child Development*, and is a member of the National Institute of Mental Health's Research Scientist Development Review Committee. In 1972, Dr. Eiduson received the Research Scientist Award from the National Institute of Mental Health which was renewed in 1977.

IRA D. GLICK.   Dr. Glick is Professor of Psychiatry at Cornell University and New York Hospital Medical Center and Associate Medical Director of Inpatient Services at the Payne Whitney Clinic. Dr. Glick has made substantial contributions in several areas of psychiatric teaching and research with a special interest in assessments of inpatient treatment programs. He is also widely known for his contributions to family therapy and with David Kessler is author of *Marital and Family Therapy*, and with Jay Hayley author of the annotated bibliography, *Family Therapy and Research*. Dr. Glick is on the editorial board of *Family Process*.

FREDERICK C. GREEN.   Dr. Green is Professor of Child Health and Development at George Washington University and Associate Director of Children's Hospital National Medical Center. Dr. Green has made fundamental contributions to the development and analysis of a broad range of health services to children. He has written and lectured widely on child advocacy, pediatric care in the ghetto and adolescents who become parents. Dr. Green was Associate Chief of the Children's Bureau in the Office of Child Development and is a Fellow of the American Academy of Pediatrics.

E. MAVIS HETHERINGTON.   Professor Hetherington is Professor of Psychology at the University of Virginia. She has contributed major studies in the fields of child development and family process and is currently engaged in significant studies of one-parent families. Her numerous publications include *Child Psychology: A Contemporary View* and *The Family and Primary Groups*. She is Editor of *Child Development*, the leading scholarly periodical in the field of developmental psychology.

HOWARD A. HOFFMAN.   Dr. Hoffman is Associate Medical Director and Director of Education of the Psychiatric Institute of Washington, D.C. He is also Associate Clinical Professor in the Department of Psychiatry and Behavioral Sciences of The George Washington University School of Medicine and Director of the Center for Continuing Education in Mental Health of the Psychiatric Institute Foundation. His clinical interests include the treatment of couples with sexual dysfunctions. Dr. Hoffman is on the Editorial Board of *Marriage and Family Review* and is president of The Psychiatric Institute Foundation.

CAROL F. HOOVER.   Dr. Hoover is on the staff of the Adult Psychiatry Branch, Intramural Research Program, National Institute of Mental Health. She has contributed valuable studies of families of schizophrenics and has been widely recognized for her work on the vexing problem of why siblings of schizophrenics don't themselves become schizophrenic. Dr. Hoover has also published fiction and poetry in several literary publications.

A. SIDNEY JOHNSON, III.   Mr. Johnson is Consultant to HEW Secretary Joseph A. Califano, Jr.; his special field is the relationship of public policy to family life. He is also Director of George Washington University's Family Impact Seminar which is composed of distinguished scholars and policymakers who review the impact of public policy on the family. Mr. Johnson was Staff Director for the Senate Subcommittee on Children and Youth which was chaired by then Senator (now Vice-President) Walter Mondale and is currently a member of the National Academy of Science's Committee on Child Development Research and Public Policy.

MELVIN L. KOHN.   Dr. Kohn is Chief, Laboratory of Socioenvironmental Studies, Intramural Research Program, National Institute of Mental Health. He is a leading authority on the relationship between social class and family process. Much of this work is summarized in his recent book, *Class and Conformity: A Study in Values*. He has also made fundamental contributions to the study of social class and schizophrenia. He is a member of the Editorial Advisory Board of the *Schizophrenia Bulletin* and is a winner of the Ernest W. Burgess Award for outstanding research from the National Council on Family Relations.

ALBERT J. MCQUEEN.   Professor McQueen is Chairman, Department of Sociology and Anthropology, Oberlin College. His numerous scholarly publications have made important contributions in the study of African urbanization and youth as well as problems of the American black family. He has been a Visiting Scientist at the National Institute for Mental Health and a member of the NIMH Research Scientist Development Review Committee.

DAVID REISS.   Dr. Reiss is Professor of Psychiatry and Behavioral Sciences and Director of the Center for Family Research, George Washington University. Dr. Reiss has conducted a broad range of studies aimed at understanding how families organize themselves to deal with their social communities; he has developed a number of laboratory methods for precisely measuring interaction process. Dr. Reiss is on the editorial boards of *Psychiatry*, the *Schizophrenia Bulletin*, the *Journal of Adolescence* and is a member of NIMH's Research Scientist Development Review Committee.

NORMAN A. SCOTCH.   Professor Scotch is Chairman, Department of Sociomedical Sciences and Community Medicine at the Boston University School of Medicine. Trained as an anthropologist, sociologist and epidemiologist, Dr. Scotch has conducted major studies of the role of socioculture factors in chronic disease with a particular interest in cardiovascular disorders and alcoholism. He is co-editor with Sol Levine of *Social Stress*, is a member of the National Advisory Council on Alcoholism and a Fellow of the American Anthropological Association.

MARVIN B. SUSSMAN. Professor Sussman is Chairman, Department of Medical Social Science and Marital Health, Bowman Gray School of Medicine and formerly held the Sealh Chamberlin Professorship in the Department of Sociology at Case Western Reserve University. He has made significant contributions in many areas of sociological research including the sociology of medicine and rehabilitation, aging, the organization of human service systems and the family. He has recently published *The Family, Bureaucracy and the Elderly* and other publications include the book *Non-traditional Family Forms in the 1970's*. Professor Sussman is Editor of *Marriage and Family Review*.

Preface

Several years ago, an anonymous donor gave a generous gift to the Department of Psychiatry and Behavioral Sciences at George Washington University School of Medicine. The donor suggested that the gift be used to support a conference on the current dilemmas of the American family and to publish its proceedings. The current chairman of the department, Jerry Wiener, formulated the initial plans for the conference with Leon Yochelson, who had been chairman at the time the gift was made. Dr. Yochelson is now Chairman of the Board of the Psychiatric Institute of the District of Columbia. These initial discussions led to a significant and sustained collaboration between the Department of Psychiatry and Behavioral Sciences at George Washington and the Psychiatric Institute in planning the conference and the present volume.

A committee was established to plan the conference. It consisted of Peter Steinglass, Associate Professor of Psychiatry, and George Cohn, Professor of Child Health and Development, both of George Washington, and Margaret Garrett, a member of the psychiatric staff at the Psychiatric Institute. The committee was jointly chaired by the editors of this volume. The committee received indispensable assistance from members of the administrative staff of the Psychiatric Institute: Al Bruce, Carol Klein, and Miriam Mathura. Margaret Schnellinger of the Center for Family Research, George Washington University, was also very helpful in all phases of planning the conference.

A successful conference was held at the Mayflower Hotel in Washington, D.C., on June 10 and 11, 1977. The major participants at the conference were the contributors to this volume. In addition, Jerry Lewis of Timberlawn Psychiatric Research Foundation, Dallas, Texas, served as a discussant. Much to our regret, we could not include his remarks in this volume because of technical difficulties.

We required additional help to complete the book reporting the proceedings of the conference. Judith Loudenback of the Center for Family Research, George Washington University, provided essential assistance as editorial assistant, as did Stephen Winter of The Psychiatric Institute Foundation.

DAVID REISS, M.D.
HOWARD HOFFMAN, M.D.

Contents

Introduction: Our Concerns about the American Family

David Reiss

This book is intended for people who work with families. More specifically, it is written for people whose work with families has concerned them and made them wonder: What's happening to the American family? People are troubled about different aspects of current family life. Included in most people's lists are:

1. The rising divorce rate
2. Continuing reports of widespread child abuse
3. Equally disturbing accounts of violence between spouses, including murder
4. Rapidly rising rates of illegitimate births

There is a second set of observations and reports about family life that are less dramatic. They are not clear indications of current distress but suggest to many how, in the future, the family might develop or decay. These observations include:

David Reiss • George Washington University, Washington, D. C.

1. The increasing number of couples who do not want children
2. The number of couples in nontraditional living and social arrangements: contract marriages, communes, homosexual marriages, and mothers who are voluntarily without husbands
3. The increasing frequency and acceptance of extramarital sexual relationships

Finally, there is a third set of concerns. These do not focus on the family *per se* but on widespread and deeply rooted social factors that seem to bruise and corrode family life. Among these concerns are:

1. The widening economic gap between black and white, rich and poor
2. Government policies or laws that interfere with or intrude upon family life
3. Deterioration of American values, which now stress materialism and consumerism

This book is a collective effort to help those with a practical concern for family life to analyze current trends. It will not provide a clear-cut answer to the question posed by the title: Is the American family dying or developing? It will provide several perspectives—drawn from history, sociology, psychology, and politics—to help analyze this fundamental question. The contributions to this book draw on three resources: carefully worked out conceptions and definitions, thorough scholarship using the best methods for collecting and analyzing data available today, and long, practical experience. Although the reader is encouraged to study each chapter carefully, we can anticipate for him the general tenor of contributions: sober optimism. The optimism has three components. First, contributors clearly emphasize the legitimacy of today's concern about the American family. Second, the contributors depict the American family in a reciprocal relationship with American society. This image emphasizes the robust properties of the American family and its capacity to innovate and change. Finally, several of our contributors outline specific remedies, changes in policies and programs, that can improve the quality of family life in America.

THE CONCERN ABOUT THE FAMILY

What would happen if the worst predictions about the family came true? Suppose (a) that marriages became transformed into nonlegal relationships of varying durations; (b) that children issued from a variety of serious and casual sexual unions and were raised in nonfamily settings; and (c) that property rights and legal responsibilities eliminated the concept of the family. We can identify two types of concern about these radical transformations: an ethical one and a practical one.

The ethical concern focuses on the quality of human life. First and foremost, it arises from a conviction that families provide their members with a richness of experience and a deep and enduring framework for absorbing and transacting with almost every aspect of daily life. This ethical concern stems from a conviction that strong and stable families are worth preserving, even cherishing, in their own right. A closely related concern is with human development. In order to become fully human, this perspective argues, we must know the physical intimacy and trust of infancy; the struggles around autonomy and the completion of childhood; the exhilaration of productive work, of sex and of a sense of ourselves in youth; and a sense of fecundity and passing on a heritage as we grow older. This trajectory of human life that makes us more human than otherwise, is impossible—this perspective argues—without a stable and strong family.

The practical concern is a different one. It arises from a preoccupation with the everyday functioning of social institutions and with the functional competence of American society as a totality. The possible demise of the family raises central anxious questions: How will things work? Who will look after the children? Who will provide the elementary training in social skills? Who will provide the stimuli for cognitive development? Who will take responsibility for the sick and for the dead? Who will initiate the needed changes in schools and neighborhoods? Who will belong to the churches and to whom are the central messages of advertising and marketing to be addressed?

A thoughtful analysis of the family is not likely to be exclusively ethical or practical. We would mourn the passing of the fam-

ily because it has, in fact, served to integrate the practical requirements of our complex society and our most ethical and humanistic
aspirations for ourselves and our fellows. Indeed, the family—as
we recognize it today—is not a self-enclosed entity. It is a per-
sonification of the strong and often practical requirements of a
complex society. It also serves to express the very human and ethical aspirations of its members. The family as an agent of practical
society falls *victim* to political and economic forces, but as the agent
of the best human hopes, it is our *agent for changing* social form and
custom.

APPROACH TO ANALYSIS

This concept, an image of reciprocity between family and society, shapes much of what follows in this book. Each contributor
recognizes and specifies certain aspects of our American society
that shape current family life and force it in specific directions. On
the other hand, the contributors perceive the family as the active
agent of the strong aspirations of its members; they see the family
as often capable of resisting social pressures and, in some cases,
transforming them. Several distinct perspectives are utilized to examine this reciprocal process.

Politics and Ideology

In Chapter 1, David Brion Davis clarifies the role the family
has had in the evolution of American democracy. His historical
analysis illumines one aspect of the reciprocal relationships between family and society. An early event was a major shift in the
political system: from European aristocracy to an experimental
American democracy. Old forms and principles, by which men's
lives were regulated, were stripped away, giving the family a more
central regulatory role. The family became bounded as a conservator of ethical traditions and, through its new function, itself
spawned fundamental social revolutions. Davis takes a long historical view. In contrast, Bernice T. Eiduson in Chapter 5 focuses on a
more immediate historical change. She begins her analysis with the
radical shift in values that occurred at the height of the Vietnam
war in the late 1960s. Inseparably intertwined with the social pro-

test of those years, particularly of the youth, was a radical revision of the basic values relating to the family. Eiduson focuses specifically on the emerging concern with genuinely intimate relationships between one individual and another and also on the relationship between all humankind and their environment. These positive values developed along with a derogation of materialism, possessiveness, and conventional conceptions of sex and age roles. She documents the influence of these new values in shaping new family forms. She is in the midst of her analysis of the impact of these new family types—social-contract marriages, communes, and single-parent families—on the development of children born into and raised by them. When her analysis is complete, she will have a grasp of the reciprocity between society and family. Does the socially engendered value of intimacy and integration—and the consequent changes in family life—produce children who then go on to make further radical revisions in society?

Stratification

All industrial and postindustrial societies consign their individual members to strata or classes. Every society has its top and its bottom. Those at the top enjoy power, prestige, and all the material fruits of their society's work. Those at the bottom get few or none of these rewards. Melvin L. Kohn, in Chapter 2, examines a fundamental source of stratification in American society: occupation. The kind of job a person does determines his position or location in our society: his income, his prestige, and his power to control others. Most important, for our analysis, is that a person's occupation also shapes the character of his family. In this respect the family is a passive responder to a fundamental condition of social life. However, as Dr. Kohn explains, recent data are now showing evidence of a more reciprocal process. The reciprocity between stratification and a resilient and resourceful family is clearly illustrated in Chapter 3 by Albert J. McQueen. He deals with the stratification of individuals by race and describes the impact on family life of American racial attitudes. The force of racism in American life has subjected the black family to enormous pressures. Departing from Frazier and Moynihan's conclusions,

McQueen clearly delineates the active response by which black families fight back and counteract the forces on them, if not transforming the forces themselves.

Child Development

Insofar as child development remains a clear function of families, the child himself becomes a major agent of the family. To some extent, the character of the child constitutes the family's response to influences from society. This is, in a large measure, Tocqueville's point as Davis restates it. The continuing viability of the American democracy depended on a particular and lasting influence of the son's relationship with his father. Our current understanding sees a more modest influence on society of a family's child-rearing practices. Nonetheless, there are many ways in which the growing child—influencing and shaping the world around him—serves as the agent of his family. He continues in this role when he becomes an adult as well. In Chapter 4, E. Mavis Hetherington, Martha Cox, and Roger Cox focus almost exclusively on this issue. Their analysis is at once alarming and encouraging. Mother-led families face stresses. Some of them cope very well, and the children develop with a full range of intellectual and social skills. Other mother-led families cope poorly, and the children grown up impaired. We might wonder, for a moment, about the children who do well. Are they simply equal in competence in a quantitative sense to children in families with a mother and a father? Or does the experience of growing up in a mother-led family equip them with a special set of attitudes and motives that in turn, lead these children to make a particular imprint on the wider society? I raised this same point in connection with findings that are just emerging from Eiduson's work: Will these children from variant family forms make their own impact on American values, perspectives, and practices?

Intervention

It is of course encouraging to view the relationship between society and the family as reciprocal. This view holds that if necessary, the family can fight back when its interests are undermined. However, McQueen's and Hetherington's chapters remind us that this is not always the case. American racism, as well as the more diffuse American pressures responsible for the rising divorce rate, have exacted an enormous toll on families. Thus, now more than at any time in recent history, there is a serious interest in government intervention in some way—particularly at the federal level—to "save the family." In Chapter 6, A. Sidney Johnson clearly shows that what we need least is a federal bureaucracy to "protect" families. Nonetheless, the current interest in some kind of intervention by the federal government leads us to ask: What does it make sense for us to do?

Sensible action requires that we know two simple things: What will happen if we do something and what will happen if we refrain from doing anything at all? Predictions of this kind are always hazardous. If we take seriously that the relationship between society and the family is reciprocal, then predictions are even more difficult. For example, if we make some change in the way society operates—as through major legislation—we cannot be certain how families will respond; they are not, according to the concept of reciprocity, just passive victims of social forces. Despite this difficulty of prediction, Marvin B. Sussman and Johnson try their hand at proposing broad-scale approaches to protect the family from some of society's most corrosive influences and to upgrade the family's capacity to cohere, endure, and maintain an active role in its reciprocal relationship with society.

Between them, Sussman and Johnson examine four different approaches to intervention. The first might be termed *policy*. This refers to a *congeries* of articulated positions that government can take vis-à-vis the family. A systematic approach to evaluating the impact of federal legislation on the family is a good example of a policy. Sussman offers another example, a public policy for developing a pool of eligible adults to provide children who need them with constant relationships with adults. A second, more specific approach is *legislation*. Johnson calls for examining legislation in detail and changing those aspects of laws that interfere with family life. As an example, he cites an income tax provision that will allow families to deduct child-care costs if the caretaker is unrelated to the family but not if, for example, the caretaker is the grandmother. A third approach is the partial *revamping of business practices*. An example here is Sussman's proposal to the insurance industry for divorce insurance. Finally, *formal education* can be an intervention. In this domain, as an instance, Sussman proposes specific and universal training for families to deal with the bureaucracies that are important to them.

It is curious, perhaps, that in a book edited by two physicians (and where clinicians are the discussants), no mention is made of clinical interventions. Perhaps the most significant clinical intervention to have an impact on families in the last two decades has been family therapy (and more recently, sexual therapy). The family therapy movement has grown prodigiously, with therapists being trained in medicine, psychology, social work, and sociology. We now have solid evidence that some forms of family therapy are very effective for some kinds of family problems. Nonetheless it's hard to evaluate the overall impact of the family therapy movement on the strength or durability of the American family. To some extent, the problem is numbers. Even now, there are too few well-trained therapists. This small number can treat only a fraction of the families who need help. To an extent, the problem in evaluating impact is that family therapy (in all its varieties) cannot, by itself, ameliorate or counteract some of the most corrosive influences on the family: racism, poverty, occupational constraints, changing values, and pernicious laws. But perhaps the most important difficulty in evaluating the overall impact of the family therapy movement is that the movement itself was designed for *cases*, not the general family. In other words, it is aimed at dealing on a case-by-case basis with specific family problems and crises in families who are perspicacious enough to ask for help. The family therapy movement cannot have an overall impact on the strength of the American family until it succeeds in bringing about widespread changes in government policy, legislation, business practices, and education. The movement, or its most articulate spokesmen, may accomplish some of this directly. More important, the family therapy movement may act to produce these broad changes through its introduction into common awareness of a new set of values, values that emphasize both the predicaments of current family life and at the same time the singular importance of the stable and durable family. We felt any such analysis of this general impact of the family therapy movement to be premature and far too speculative. It is well to observe that it is only in the last two decades that the broad-scale social impact of the clinical interventions of a prior age—pyschoanalysis—has been effectively described. Perhaps the best examples are those of Trilling (1955) and Rieff (1966).

This discussion of clinical interventions brings us full circle, back to the concern of those who are practically engaged in the welfare of specific families. This book will not offer you technical tools for your daily work. We hope, however, that it will provide you with analytic perspectives that will help orient you and, as we have already anticipated, will do so with a sense of sober optimism.

REFERENCES

Rieff, P. *The Triumph of the Therapists*. New York: Harper & Row, 1966.
Trilling, L. *Freud and the Crisis of Our Culture*. Boston: Beacon Press, 1955.

I

American Historical Development and the Family

1

The American Family and Boundaries in Historical Perspective

David Brion Davis

TOCQUEVILLE AS A FATHER FIGURE FOR AMERICAN CULTURE

Since the 1840s, and especially since the 1940s, many Americans have turned almost compulsively to Alexis de Tocqueville's *Democracy in America* as the authoritative fountainhead of historical self-understanding. Tocqueville's prestige, which admittedly has had its ups and downs, owes something to his refreshing departure from the English empirical tradition, or to what the 11th edition of *The Encyclopaedia Britannica* calls his "excess of the deductive spirit." As one of the founders of classical sociology, he was also the first European to apply systematic social theory to American institutions. But I suspect that Tocqueville's appeal to Americans rests mainly on his ability to portray the United States as the vanguard

David Brion Davis • Yale University, New Haven, Connecticut.

of a long-term and irresistible movement toward democratization and modernization and to explain this revolutionary process by continual contrast with a premodern and aristocratic European tradition. Though Tocqueville was only 26 when he toured America in 1831, he soon became a cultural father-figure. For not only did he serve as interpreter and spokesman for the aristocratic past, but he assumed the role of an enlightened parent who accepts the inevitability of growth and change, who wants to understand, and who offers counsel designed to mitigate the excesses and hazards of freedom.

While Tocqueville could not conceal his nostalgia for the ordered past, he frankly acknowledged the justice of democracy and the moral benefits of expanding equality. For Americans, his image of fixed and hierarchical boundaries was not a goal to be recovered but a reference point for measuring their own anxieties and aspirations, a peak of the Old World still in view to confident but giddy voyagers. Thus Catharine Beecher, whose 1843 *Treatise on Domestic Economy* has been likened in impact to Dr. Benjamin Spock's *Baby and Child Care*, repeatedly invoked Tocqueville to justify the preeminent mission of American mothers. Tocqueville had shown, she wrote, that in aristocratic lands "all ranks and classes are fixed in a given position, and each person is educated for a particular sphere and style of living." The price of American equality had been a loss of direction and coherence. In a land of rampant change, fluidity, and self-seeking individualism, only the family remained as a stabilizing and regenerating force. Because Catharine Beecher agreed with Tocqueville that in America the shaping of character was "committed mainly to the female hand," she concluded that "to American women, more than to any others on earth, is committed the exalted privilege of extending over the world those blessed influences, which are to renovate degraded man."[1] Women, in other words, by accepting the duties of their household sphere, were the providential antidote for the ills of the modern world.

THE ILLUSION OF A GOLDEN AGE OF FAMILY STABILITY

As we shall see later on, both Tocqueville and Beecher, to say nothing of countless others, assigned to the family burdens that

---

[1]Quoted in Sklar (1973).

had once been dispersed through less functionally differentiated societies. But the Tocquevillean device of measuring the present against a more stable and ordered past—or what Ferdinand Tönnies would eventually describe as a *Gesellschaft* versus a *Gemeinschaft* society—also suggested a certain fall from grace, a decline from a golden era of reciprocal rights and duties, of organic and harmonious unity, when everyone honored the elderly and when children derived wisdom and joy from grandparents, aunts, uncles, and other kin. In fact, it would appear that this idyllic image is a fantasy of modern times, that is, of the 18th and especially the 19th centuries, when the family acquired a new symbolic importance as the embodiment of the past and as a refuge from and antidote to a world of bewildering change. It was not accidental that the idealization of the bourgeois family emerged in close association with the literary idealization of Laplanders, American Indians, South Sea islanders, and other allegedly primitive and innocent peoples. The two trends were closely tied to the so-called discovery of childhood as a distinct, almost continental sphere of being (Victor Hugo is credited with saying, "Christopher Columbus only discovered America: I have discovered the child"). These various quests for models of simplicity, spontaneity, and wholeness may well have been rooted in deep and perhaps universal psychological longings for a world without boundaries, for a global union of self with all being, or what Freud termed an "oceanic feeling." One may theorize that such universal human longings were cast adrift, so to speak, by the diminishing credibility of myth, ritual, magic, and belief in the supernatural, and were further accentuated by the growing rationalization and differentiation of life. But the immediate point at issue is the reality of an earlier golden age, not the reasons for its discovery or idealization.

During the past decade, the history of the family has undergone exciting and explosive growth. Because of the diversity and complexity of the evidence and because our knowledge is expanding so rapidly, it is extremely hazardous to generalize about broad trends that include various social classes and nationalities and that extend over several centuries. Nevertheless, there is good reason to doubt theories of abrupt change in the character and viability of the family. In a recent and illuminating review of the literature from various scientific disciplines, including anthropology,

biosociology, and neuroendocrinology, Alice S. Rossi (1977) stresses the central parenting bond between mothers and small children, a bond whose genetic and evolutionary antecedents derive
from the hunting-and-gathering societies that account for over
90% of human history. Quite apart from this biosocial continuity,
historians have discovered that the supposedly "modern family"—that is, the nuclear family of two parents and their children living in an independent, private household—became predominant in western Europe several centuries before industrializa-
tion. At an early period, Western Europe also diverged from the
rest of the world in a number of other respects: in the late age at
which females married (mid- to late 20s); in the high proportion of
women who never married or bore children; in a "planning mental-
ity" geared to long-term advances in family wealth and status; and
in the appearance, at least in 18th-century France, of conscious and
effective efforts at contraception. It has been convincingly argued
that these modern familial patterns may have been more the cause
than the consequence of the process of "modernization."

Even more forceful objections have been raised against the picture of decline from a previous golden age of familial stability and
harmony.[2] Because of heavy mortality, three-generation households were always rare until the late 19th century. (While growing
up, the majority of children lost one parent.) By present-day
standards, the proportion of people over 60 in any population was
very small—in 17th-century England, no more than 8%—and the
economic burden of supporting the elderly was correspondingly
light. Nevertheless, respect for the elderly depended in large part
on their retention and control of property. The elderly who had no
property to be coveted by heirs or to be legally deeded in exchange
for lifetime support might find themselves reduced to beggary or to
a marginal existence on parish welfare.

The treatment of children also dramatizes the failure of religious and ethical norms to protect the powerless, although such
norms were far more effective in America than in Europe. It now
appears that a massive decline in infanticide was a major contributor to western Europe's population explosion of the 18th century (McKeown, 1977; Stone, 1974). The earlier prevalence of infan-

---

[2]See Stone (1977) and Thomas (1976). But for a contrary view see Fischer (1977).

ticide probably encouraged a degree of aggression and "careless-ness" toward small children in general. In this respect, Christian nations may well have differed psychologically from societies where infanticide was a legal and unquestioned means of popula-tion control. Since in Europe infanticide was legally defined as murder, its practice, even as an "accepted custom," had to take the form of an "accident"—suffocation, "lying-over" in bed, neglect, or beating. Archaeological evidence indicates an appalling number of infants and small children who died from physical violence or who were maimed in various ways to enhance their success as beggars (as is still done in India today).

There can also be little doubt that children were often indefen-sible targets for adult sexual gratification. This is a subject about which we know too little, and a subject to which we shall return when considering the prudery of the "Victorian" family. Philippe Ariès (1965), who was the first to explore the question in some depth, warns us against imposing our own values on the 17th-century aristocrats, and their servants, who played sexual games with toddlers and who toyed with the penises of princes. Ariès and other scholars have shown that the very idea of children's sex-ual innocence is a relatively recent development. Yet this early lack of inhibition should not be confused with modern ideals—some would say fantasies—of an unrepressed and Eden-like state of na-ture. Ariès's own evidence does not suggest sexual education and initiation, as in many premodern societies, but rather the impulsive exploitation of very young children who were hardly yet consid-ered human beings. Admittedly such evidence is confined to the aristocracy and can hardly be taken as representative. But future research may well confirm Ariès's contention that "playing with children's privy parts formed part of a widespread tradition, which is still operative in Moslem circles."

There are still other grounds for questioning the idyllic image of preindustrial family nurture. In Europe and even in colonial America, the newborn infants of well-to-do families were com-monly "put out" to wet nurses for a period of a year or two. In France, where the custom was especially prevalent, nurses were recruited from among the most debased and exploited class, and infant mortality was correspondingly high. Deprivation was the hallmark of such absentee and unsupervised child care—many wet nurses had just weaned their own children and because of a low-energy diet thus had a diminished supply of milk. And of course the children faced the conscious traumas of separation at age 1 or 2. For upper-class children, by the early modern period, adjustment to the natal home might be followed in a few years by banishment to a boarding school of some kind. Less affluent families regularly sent out children as apprentices or servants, often as early as age 7, at which point guidance and discipline shifted from parents to a nonkin family.

One must also picture the plight of the colonial New England mother who regularly gave birth to a new child every two years while cooking, spinning, weaving, baking, pickling and preserving, making candles and soap, bleaching, dyeing, cleaning, laundering, and gardening. Unlike many Europeans, New Englanders did not immobilize toddlers in tight swaddling clothes, and it is not surprising that many children were burned by fire, fell into wells, or were the victims of other accidents indicating a lack of adult supervision. In both Europe and America, the desperation of parents, nursemaids, and others charged with child care is suggested by the widespread use of opium, alcohol, and other sedating narcotics. But as John Demos (1971) has written, New Englanders began exercising their principal control in the child's second year, when they resolved to break or beat down inherent "willfulness" as soon as it began to appear. The Puritans were not alone in conceiving this to be "*the central task* of parenthood," or to regard it, as Demos says, "as involving a direct confrontation with 'original sin.' " In this battle against Satan, adults could employ an imposing arsenal that ranged from the exhibition of corpses to folktales of castration, witchcraft, and abandonment.

One can easily distort historical reality by dwelling on the horrors of the past, and at this point I should stress that the preindustrial family performed a variety of economic, religious, educational, and welfare functions that were eventually assumed, for good or ill, by other institutions. Although the so-called modern family appeared in *form* at a surprisingly early date—that is, in the formal structure of conjugal units living in independent households—the *functions* of the family continued to be varied and diffuse, particularly in times of emergency. Indeed, there is a certain irony in the fact that the stunning uprootings and dislocations of the early in-dustrial era encouraged a return among laboring families to earlier patterns of mutual aid based on extended kinship and ethnicity. At a time when laissez-faire dogma discredited every surviving form of parish or state paternalism, the family filled the gap as the only agency of welfare and mutual assistance. This phenomenon, with many cultural variations, was particularly striking among black freedmen in the post-Civil War South and among industrial immi-grants, who evolved new forms of extended family dependence as well as mutual-aid societies based on shared religious or ethnic background. But the history of social welfare legislation, carrying with it appalling records of the human costs of early industrializa-tion, proves that the resiliency of the family could be no substitute for an abdication of social and political responsibility. Nor was the family then or earlier a haven of security, order, and mutual self-fulfillment.

I have quite deliberately applied our own standards to the past, in order to highlight the opposite fallacy of idealizing the past and of forgetting that the past two centuries have witnessed a momentous change in moral sensibility. This change has been obscured by continuing laments over the decline and impending disintegration of the family. For over three centuries, ministers, moral philosophers, and their modern counterparts have been warning that children are becoming more contemptuous of author-ity, parents more irresponsible, wives more selfishly independent, families more fragile and rootless. Seldom have these prophets of doom taken note of the subtle transformation in functions and ex-pectations assigned to the family. As we have already briefly noted, the family has been stripped of many of its traditional functions; and as we shall see, it has acquired awesome new re-sponsibilities for ensuring individual happiness and social salva-tion. As Philippe Ariès (1977) has recently pointed out, I think with some exaggeration, the feelings and emotions that were once dif-fused among various natural and supernatural objects, including saints, friends, horses, and gardens, have become "focused en-tirely within the immediate family. The couple and their children became the objects of a passionate and exclusive love that tran-scended even death." What both Ariès and less knowledgeable lamenters tend to forget is that the 18th century marked the first upsurge in history, and then in a few select societies in Western Europe and America, of a *collective* concern for the fate of helpless and vulnerable humans—of children, the aged, slaves, the insane, the physically handicapped—for a fate, that is, apart from that determined by family or kin. And by post-18th-century standards, family and kin have seldom had the resources or goodwill to provide the warmth and security imagined in nostalgic retrospect. The rather priggish Victorians who ventured into what we now call the Third World, and who recoiled at the sight of infanticide, child maiming, child prostitution and enslavement, widow burning, and all the rest, would have been similarly shocked if a time machine had transported them backward in their own societies two centuries or more.

TOCQUEVILLE AND THE SIGNIFICANCE OF BOUNDARIES IN A DEMOCRATIC SOCIETY

Having questioned widespread assumptions about the nature of the preindustrial family, I return to Tocqueville for two reasons. He presented a perceptive if highly theoretical analysis of the American family before the onset of significant industrialization but after a half-century or so of profound socioeconomic change, especially in the status and economic role of women. Tocqueville also sensed that the family somehow exemplifies the restructuring and redefinition of boundaries in a democratic society. His major concern, it should be stressed, was how the irresistible movement toward equality could be "civilized," so to speak, by a voluntary acceptance of various restraints and limits. And such voluntary acceptance, in Tocqueville's view, depended on the nature and the future of the family.

Tocqueville's great insight, which was in no way diminished by his bias or his factual errors, is that the historical drift toward equality impinges on boundaries of every kind: psychological, social, political, religious, and territorial. Not only is the family itself shaped by the trend toward equality of condition, but it is at once a key source, amplifier, and stabilizer of egalitarian aspirations, of individualism, and of the restless pursuit of success. With a view to the themes I wish to develop in the rest of this paper, Tocqueville's paradigm provides us with useful concepts for relating sex-role changes to the larger process of modernization; for understanding the new burdens and expectations that were just beginning to be placed on the American family; and for identifying the long-term costs that were associated with the elevation of women and with a new parental concern over "child development."

The significance of interrelated boundaries can be seen in the way that Tocqueville first introduced the subject of the American family while discussing the stabilizing influence of religion on democracy. Because Americans universally accepted a generalized and nonsectarian Christian morality, Tocqueville (1955) wrote:

> every principle of the moral world is fixed and determinate, although the political world is abandoned to the debates and experiments of men. Thus the human mind is never left to wander over a boundless field; and whatever may be its pretensions, it is checked from time to time by barriers that it cannot surmount. Before it can innovate, certain primary principles are laid down, and the boldest conceptions are subjected to certain forms which retard and stop their completion.

In other words, religion provides an outer rim of unquestioned rules, limits and assumptions, but within this circle custom, tradition, and political authority are all open to debate and transformation. But how does a vague and generalized religion retain such influence, particularly when, as Tocqueville admits, it is "often unable to restrain man from the numberless temptations which chance offers; nor can it check that passion for gain which everything contributes to arouse"? Certainly the force of religion cannot be explained by the Church as an institution, since Christianity in America is fragmented into scores of competing sects; nor can it be explained by the power and patronage of the clergy, who "keep aloof from parties and from public affairs."

Tocqueville concluded, almost casually at this point, that Christian morality (which he rather carelessly equated with "religion") is rooted in the home and is somehow connected with the ideal of "conjugal happiness." Significantly he also claimed that in Europe, rebellions against authority—the kind of sweeping, nihilistic rebellions that he most feared—originate in the home, in the "irregularities of domestic life." Such "irregularities" appear to include both an Oedipal struggle against paternal despotism and a cavalier regard for marital fidelity. As a result of such contempt for what Tocqueville called "the natural bonds and legitimate pleasures" of the home, Europeans were inclined to *transfer* rebellions and lawless impulses outward:

> Agitated by the tumultuous passions that frequently disturb his dwelling, the European is galled by the obedience which the legislative powers of the state exact. But when the American retires from the turmoil of public life to the bosom of his family, he finds in it the image of order and of peace. . . . While the European endeavors to forget his domestic troubles by agitating society, the American derives from his home that love of order which he afterwards carries with him into public affairs.

No doubt this contrast is overdrawn, but it offers intriguing possibilities for psychohistorical research, and, perhaps more important, it reflects an emerging ideology that assigned new psychological responsibilities to the middle-class home and to American women. Almost in passing, Tocqueville observed, still regarding religion, that "its influence over the mind of woman is supreme, and women are the protectors of morals." Many chapters later he returned to this theme and admitted that if asked:

> Now that I am drawing to the close of this work, in which I have spoken of so many important things done by the Americans, to what the singular prosperity and growing strength of that people ought mainly to be attributed, I should reply: To the superiority of their women.

This was the message, as I have already briefly noted, that was seized upon by Catharine Beecher and other champions of the so-called cult of domesticity. Recent studies have shown that during the first decades of the 19th century, the Protestant clergy devoted increasing attention to women, who formed a growing majority of their congregations. The mission prescribed for women was not one of passive holiness or spirituality but of active work, mainly as mothers and wives, in saving masculine America from the effects of materialism, dollar worship, and competitive struggle. Historians have only recently begun to see that this evangelical enlistment of middle-class women had vast multiplied effects extending beyond the household into virtually every movement for social reform.[3]

---

[3]See Cott (1977).

Before we pursue this momentous division between masculine materialism and feminine idealism, it is important to summarize Tocqueville's more detailed diagnosis of the democratic family. First, like most European observers, Tocqueville was struck by the weakness of paternal authority, which did not reside in a kinglike *office* but depended rather on "natural affection" and on a temporary advantage of age and experience. The paternal bond was thus informal, permissive, and premised on the early independence of sons:

> The father foresees the limits of his authority long beforehand, and when the time arrives, he surrenders it without a struggle; the son looks forward to the exact period at which he will be his own master, and he enters upon his freedom without precipitation and without effort.

Similarly, whereas siblings in an aristocratic family were bound together by tangible family interests, jealousies, and obligations, democracy "divides their inheritance, but allows their hearts and minds to unite." In short, democracy "destroys or obscures almost all the old conventional rules of society" but at the same time tightens the bonds of natural feeling and sympathy.

A second characteristic, which again drew comment from numerous Europeans, was the seeming paradox that "in America the independence of woman is irrecoverably lost in the bonds of matrimony. If an unmarried woman is less constrained there than elsewhere, a wife is subjected to stricter obligations." Parents, instead of trying to shelter their daughters from the vices and temptations of the world, trained them to rely on their own strength of character. Yet upon marriage the self-reliant American woman accepted "a constant sacrifice of her pleasures to her duties" and a role defined by public opinion that "carefully circumscribes woman within the narrow circle of domestic interests and duties and forbids her to step beyond it." We now know that many young women experienced what one historian has termed a "marriage trauma" while either making or recoiling from this difficult transition, but the evidence suggests that for the middle class, Tocqueville's description was essentially correct.

It must be stressed, however, that Tocqueville was describing "ideal types" that cannot be taken too literally but that point to important historical trends. For example, in discussing what he

termed "the equality of the sexes" in America, Tocqueville over-
looked, or as a male simply took for granted, the actual inequalities
that seem so shocking in retrospect: the minimal educational opportunities for girls; the limited access to employment outside the
home, and pay-scales of one-third that for males in similar jobs; the
denial of political rights and, in the case of married women, of the
right to own property, to make contracts, or even to make a will.
Yet there can be no doubt that the long-range trends favored
greater sexual equality. Moreover, as Tocqueville acutely saw, the
immediate path toward equality did not imply similarity or a blur-
ring of gender boundaries:

> In no country has such constant care been taken as in America to trace
> two clearly distinct lines of action for the two sexes and to make them
> keep pace one with the other, but in two pathways that are always
> different.

The familial patterns that Tocqueville described were part of a
profound process of societal change, a process that can be vaguely
termed *modernization* but that can best be understood as the de-
struction and compensatory reconstruction of boundaries of every
kind. This leveling and reordering process had a long history and
in 1831 was probably still confined, in its more extreme forms, to
the Northeastern middle class. While industrialization would eventually accelerate the rate of change, the original and critical transformation of the American family was intermeshed with earlier historical developments that can be only briefly enumerated here.

First was the gradual imposing of voluntary limits on human
reproduction. During the colonial period, a plentiful food supply
had helped to remove the ancient and lethal balance between fertility and mortality, leading to a population that continued to double
every 20-odd years. Although a similar growth rate persisted well
into the 19th century, fertility began declining in the late 18th century. By the second quarter of the 19th century, long before any
significant use of contraceptives, middle-class American women
were consciously limiting the size of their families, the space between children, and the age at which their childbearing would
cease. The population as a whole was gradually aging, and in the
Northeast, adult women increasingly outnumbered adult men.

A second set of developments was the so-called transportation revolution, increased agricultural productivity and specialization, the growth of towns and seaports, and the spread of a market-oriented economy. For the family, these changes brought a rapid decline in economic self-sufficiency and in household industries, such as spinning and weaving, where women and girls had predominated. The need for cash income for a time involved many households in various putting-out systems of production. But increasingly the place of work was removed some distance from the home, whether for a male "breadwinner" or for daughters who lived as boarders while teaching school or working in textile mills. Housework, to be sure, was hardly less arduous than before, but in male eyes, the chores of mothers and wives lost the status of "work" when divorced from household industries. Psychologically the home became radically separated from the alienating and dehumanizing pressures of "modern work," that is, work defined by market forces and by impersonal standards of maximum efficiency, production, and profit. Though growing numbers of women, especially young unmarried women, worked temporarily as wage earners, this simply reinforced the dominant expectation that women's true and permanent "sphere" was the home, a recuperative oasis in the midst of a warlike jungle of competitive self-seeking. Significantly, the image of the home as a preserver of the organic and cooperative values of the past presupposed traditional rhythms of time and responsibility. For example, if the wife was blessed with the supposedly transcendent calling of motherhood and household management, her labor was task-oriented, irregular, and unsupervised, yet unending. Unless exempted by illness, she was on call 24 hours a day.

It should be stressed, however, that this concept of a segregated domestic sphere was part of a broader and truly revolutionary redefinition of woman's role. Precisely because middle-class women could claim to be untainted by the amoral expediency of the business world, they could demand a public hearing in movements like temperance and antislavery. There were complex connections, which cannot be spelled out here, between the cult of domesticity and the emergence, by the 1830s, of militant feminism as a new force in history.[4] The discovery of new modes of feminine self-assertion was also a response to political democratization, since the enfranchisement of adult white males had the effect of accentuating barriers of race and sex in the name of a more open and egalitarian society. Finally, women turned ideals of domesticity against the proliferating clubs, fraternal lodges, and barrooms that provided for segregated male recreation. This long campaign, finally and decisively aided by the 20th-century flight to the suburbs, can be viewed as part of the process of converting the family into a consumer society's primary unit of consumption.

To sum up, a focus on changing and interrelated boundaries suggests that the 19th-century family was as much the agent as the victim of modernization. The wife who made a self-conscious vocation of motherhood was as much a force for change as the female employee of a textile mill. The hallmarks of modernization were increasing differentiation of roles and specialization of functions. By the mid-19th century, in addition to the all-important division between home and work, there was a far more rigid and uniform timing of life stages: the timing of childbearing in marriage; of the duration of formal schooling; of the entrance into employment; of the age at marriage; and of a lengthening period when spouses could expect to live together after children had ceased to be responsibilities. In the economic world, such differentiation and specialization were still primarily geared to increased productivity. The family, having lost its main functions as a productive unit, became the key institution for adapting to socioeconomic change. Hence its burdens were increasingly psychological and ideological, and in serving these needs, it became increasingly oriented to various ritualized forms of consumption.

EVALUATION AND OVERVIEW

The American family failed to meet Tocqueville's expectations as a source of order and morality counterbalancing the rampant self-seeking of the modern capitalist world. Tocqueville was blind to the material forces of history and could not foresee the commercial exploitation of what he termed "natural feeling and sym-

---

[4]See Cott (1977), which traces these connections and brilliantly elaborates the themes I have summarized in the preceding paragraph.

pathy." But before we look at long-term costs, it is important to stress the genuine gains, when judged by democratic and humanitarian standards, that accompanied the modernizing conquest and redefinition of familial boundaries.

Only a confirmed misogynist could deplore the growing insistence of American women that marriage not necessitate childbirth at two-year intervals until menopause; and only a confirmed misanthrope could express misgivings over the decline in infant mortality that, coupled with fewer childbirths, allowed parents to invest more love with less risk in each individual child. Yet these triumphs over previous natural limits cannot be applauded as isolated demographic phenomena; they carried far-reaching implications. Parents who became deeply involved in the welfare and the future of each child easily gravitated toward the modern ideal of developing the best potentialities of every human being. This humanitarian ideal probably originated in changing familial relations and in a growing tendency to perceive children as embodiments not of Original Sin but of innocence and of inclinations that could be trained for good or ill. Even in the 18th century, the task of parenthood gradually shifted toward sympathetic nurture and guidance, although the latter by no means excluded physical punishment as a means of cultivating self-discipline.

The sexual repression and family prudery that we loosely term Victorian were closely related to the modern concern with child development, self-discipline, and social progress. In this respect, it is worth emphasizing that efforts to stamp out children's masturbation became a cultural obsession only in the 19th century. Feminist historians have recently suggested that middle-class wives achieved a degree of independence and moral autonomy by insisting on periods of sexual abstinence, and it can be similarly argued, though much more evidence is needed, that efforts to negate and deny childhood sexuality originated as means of protecting children from molestation by adults. In other words, children could be defined as innocent only if they were first desexualized and hence kept from being Satan's tempters to various servants and adult relatives in the same household. Emphasis on self-discipline and avoidance of "bad habits" also tied in with the all-important temperance movement, which exemplified the ideal of nurturing individual character within scrupulously respected family boundaries.

For not only was alcohol blamed for child abuse and molestation, for male sexual tyranny over wives, and for the violation of domestic peace and the squandering of family resources, but the child who had avoided bad habits of so-called self-abuse was expected to have the power of will to resist all forms of intemperance.

Unfortunately, the moral boundaries ensuring wives and children a degree of protection and self-determination depended on symbols of internalized discipline and self-denial that entailed heavy costs. One such cost was the classic 19th-century neurosis. And here let me note that Freud, whose initial discoveries appeared toward the end of this walling-off and walling-in era, concluded with astonishment that the childhood seduction by fathers of his female patients was an intrapsychic event, or fantasy. Yet historically it may have been a more common objective event, and one may speak hypothetically of a progression from external family interactions toward subjective or intrapsychic family relations, a progression coinciding roughly with the appearance of truly private rooms not only for sleeping parents but for their children. The development of the Victorian model of superego, Oedipal conflict, and neurosis may thus be conceived as a psychological variant on the historical process of differentiation and specialization.

A second long-term cost came due with the rebellion, to which Freud unintentionally contributed, against all forms of repression. Upholders of what Tocqueville had called "Christian morality" warned that the family could never survive the defeat of national prohibition, which had culminated a century of temperance agitation, to say nothing of the license increasingly given to sexual relations outside the bond of marriage. And here we should note the peculiar fact that while the American family had long been deprived of various supportive institutions, such as those found in more traditional, homogeneous societies, it continued to receive compensatory support from incredibly detailed and authoritarian laws governing sexual behavior. Indeed, although Americans proclaimed themselves the freest people on earth, they passively assumed that state legislators could apply distinctions of legality and illegality even to the sexual conduct of husbands and wives within the privacy of a bedroom. The absurdity of such laws became apparent only with the rebellion against so-called puritanical or

Victorian values. No less important was the critical redefinition of liberty, in roughly the mid-20th century, from the notion of being as free as everyone else to the notion of being free to be different (Potter, 1976). This revolutionary change had, significantly, little effect on the laws, which remained, in the eyes of progressives, as mere sops to reactionary tastes. For progressives, the failure to suppress the drinking of alcohol simply dramatized the inability of puritanical moralists to enforce their own standards on the vibrant, fun-loving community at large. What went unnoticed was the discrediting of law and the further privatizing of the family as an institution supposedly immune from society's legitimate interests.

The American family has shown a remarkable strength, continuity, and adaptability as it has acquired an overload of expectation, much like a lonely life raft after the sinking of an ocean liner. It was helped, at least until the 1930s, by the example of immigrants, blacks, and other subcultures for whom the family could never appear as a therapeutic and easily expendable luxury. Nevertheless, it seems undeniable that the family's increasing isolation has weakened its capacity to function as a unit or to resist the manipulation of its members by various outside forces, such as advertising and peer-group norms. The continuing need for conjugal unions, as a testing ground for intimacy and authentic feeling, has been partly met by less demanding and more narcissistic sexual relationships. The ethic of self-centered liberation and fulfillment may not be compatible with parenting that extends beyond the joyful production of babies or with the notion of a sanctuary where people may safely recuperate or express feelings that would have no hearing in the outside world. But as I have repeatedly tried to emphasize, the problems of the family can never be isolated from the problems of society at large. In fact, the failures of the middle-class family are less troubling than the realization that it has performed its extraordinary functions mainly as a unit of extravagant and wasteful consumption and that it is this economic role that has made the family indispensable for a capitalist society.

This point seemed innocuous so long as intelligent planners could count on increasing per capita product, and hence per capita consumption, as the solution for human ills. Only a few years ago, the American family appeared to be a model of innocent and carefree abundance—with its separate rooms for each child, its educational toys and travel, its music and swimming lessons, its grassy spaces and child-centered outlook—a model that any society could achieve with sufficient productivity and scientific modernization. The vision lives on, despite our dark glimpses of a new reality principle.[5] But without debating the options between various risky and even terrifying futures, I suggest that the crisis over energy and ecology sheds stunning light on the contradictions and sustaining dynamics of the modern middle-class family.

The contradictions arise from the history we have been examining. In a rootless democratic society, the home became the only agent of conservation, the only symbol of a tangible past, the only place where achievement gave way to bonds of affection and loyalty and where people were loved because of who they were. Yet increasingly the family as conservator merged with the family as consumer. During the 19th century, the affluent home could be maintained as a conservatory of culture only by forays into the outside world—forays to the theater and the art museum, forays of tourism and adventure that stocked the household with souvenirs, bric-a-brac, *objets d'art*, and "relics"—the latter term having been originally applied to the holy artifacts that Crusaders brought back from Palestine. Such trophies became embarrassingly standardized; the game of showmanship became more subtle. Twentieth-century homes are far less cluttered than were the family museums of the Victorian era. Yet the message, and the contradiction, is much the same. The family homestead, as an embodiment of historic values and as a refuge from the commercial world, has depended on a symbolic display that can be purchased only in the commercial market. What has changed is that the market has become dependent on encouraging such display.

As advertisers and developers have long sensed, the disruptions of the modern world have greatly enhanced the appeal of supposedly premodern family solidarity—a solidarity that can be achieved, in fantasy, in suburban track houses that become ranches by virtue of wagon wheels flanking a driveway, or plantations by virtue of small statues of liveried slave boys reaching out to take your horse, or New England manses by virtue of stone walls and

---

[5]See Hirsch (1976) and Heilbroner (1977).

fake wells. These one-acre manors are commonly fenced off, are protected by fierce dogs or electronic alarms, and, in keeping with the old American adage that "a man's house is his castle," contain sufficient arms to repel any invader. They also contain washers, dryers, disposals, dishwashers, air-conditioning, color television, gasoline-powered lawnmowers, and central heating. They have been designed, in fact, to consume a maximum amount of the world's resources and energy and thus to create more jobs for the future occupants of similar but improved houses. No one has begun to calculate the overhead cost of establishing from scratch, for each generation, such independent households equipped with appliances and ready for the photo albums, movies, and vacation souvenirs that record the history of an American family.

Let me stress that I am not questioning the benefits of technology or mourning over our undeniable weakness, as human beings, for material possessions. The point is that the American family, which was long seen as a moral counterforce to competitive individualism, has for that very reason become vulnerable to commercial exploitation. For it has been in the name of strengthening the family and supporting its virtues and independence that public policy, since World War II, has encouraged the exodus from cities, the construction of commuter highways, the purchase of automobiles, and the so-called development, at enormous private profit, of track housing and suburban shopping centers—all of which have made the middle-class family an outlet for appalling waste and duplicated effort.

The ultimate question for us is the effect of such exploitation on the family. And here one must recall that the premodern family was governed by sordid calculations regarding doweries, family alliances, and the transmission or consolidation of power. It is surely no more sordid for children to aspire, in their own future families, to outstrip their parents as extravagant consumers. Whereas the authority of fathers was once reinforced by landed property or productive skills that could be transmitted to sons, what is transmitted now is a style and identity as consumer—a transmission and training marked off by such rituals as Christmas and family vacations, much as the child's progression as consumer is marked off by his readiness for a tricycle, a swing set, a television, or a car.

For several millennia at least the family has served as humanity's nurturing and socializing institution, even among slaves and similarly dispossessed peoples. Only within the past century or so, and only within privileged groups confined largely to the Western world, has the family become the combustion engine for an economic system that is rapidly burning itself out, a system characterized, in Jacob Epstein's (1977) words, by declining profits and by

> the constantly rising cost of an extravagant standard of living whose average quality was steadily declining—a decline that could be seen in the collapse of neighborhoods, the wreckage of the natural environment, and the compulsive production of generally useless and increasingly expensive goods while real needs went unmet.

Possibly the family will be strengthened as we ultimately become forced to adopt less wasteful and more cooperative modes of consuming limited resources. One suspects that some of the narcissistic pressures that have weakened the family, such as an apparently growing incapacity to express deep or sustained affect toward people and things, may be related to our instantly disposable and replaceable consumerism. Through premodern, modern, and postmodern periods, the family has shown great resiliency, especially in its parenting role; its role as a sanctuary for self-indulgence may thus be a briefly transitory stage in an enduring history.

REFERENCES

Ariès, P. Centuries of Childhood: A Social History of Family Life, tr. by Robert Baldick. New York: Vintage Books, 1965, pp. 100-127.

Ariès, P. The family and the city, Daedalus, 1977 (Spring), p. 229.

Cott, N. F. Bonds of Womanhood: Woman's Sphere in New England, 1780-1835 New Haven, Conn.: Yale University Press, 1977.

Demos, J. Developmental perspectives on the history of childhood. In T. K. Rabb & R. I. Rotberg (Eds.), The Family in History: Interdisciplinary Essays. New York: Harper Torchbooks, 1971, pp. 132-133.

Epstein, J. Capitalism and socialism: Declining returns. New York Review of Books, 1977 (February 17), p. 37.

Fischer, D. H. Growing Old in America. New York: Oxford University Press, 1977.

Heilbroner, R. L. The false promise of growth. New York Review of Books, 1977 (March 3), pp. 10-12.

Hirsch, F. Social Limits to Growth. Cambridge, Mass.: Harvard University Press, 1976.

McKeown, T. The Modern Rise of Population. New York: Academic Press, 1977.

Potter, D. M. *Freedom and Its Limitations in American Life*, ed. by D. E. Fehrenbacher. Stanford, Calif.: Stanford University Press, 1976.

Rossi, A. S. A biosocial perspective on parenting. *Daedalus*, 1977 (Spring), pp. 1-31.

Sklar, K. K. *Catharine Beecher: A Study in American Domesticity.* New Haven, Conn.: Yale University Press, 1973, p. 159.

Stone, L. Massacre of the innocents. *New York Review of Books*, 1974 (November 14), pp. 25-31.

Stone, L. Walking over grandma. *New York Review of Books*, 1977 (May 12), pp. 1-16.

Thomas, K. Age and authority in early modern England. *Proceedings of the British Academy*, 62, 1976.

Tocqueville, A. de. *Democracy in America*, tr. by Henry Reeve, rev. by Francis Bowen, ed. by Phillips Bradley. New York: Vintage Books, 1955, 2 volumes.

Practical Considerations

Carol F. Hoover

Professor Davis's paper supplies us with a philosophical perspective on the American family over time, which is of more than academic interest to people like ourselves who deal with current families in a variety of contexts. Real families act the way they do because they regard themselves in certain ways, as well as thinking of themselves in a particular mode based on the way they act. Moreover, a remarkable preoccupation with virtue and morality (perhaps an outgrowth of the "momentous change" in moral sensibility in the past two centuries, to which Dr. Davis refers) seems a special characteristic of the American family.

In clinical work, I have been impressed with how deeply these considerations are held and how self-evident they seem to the holders; families operate on a set of built-in moral assumptions that the members imagine, without even formulating the notion clearly, must be shared by therapists, and when families believe they *ought* to behave in certain ways, that is how they see themselves behaving. So if everyone has acted as he or she ought, then how does it

Carol F. Hoover • National Institute of Mental Health, Bethesda, Maryland.

happen that troubles have descended upon the family—or at least upon one of its members?

In general, there is an insistence that only one of these family members is showing signs of difficulty or aberration. A husband blames his wife, the wife her husband, the child his parents, the parents a child. Thus not only personal guilt is avoided but a kind of family guilt or taint that would place the entire group outside community norms and ethical or religious approval. People in the family will beg, plead with a therapist to point out precisely who is at fault, to identify an individual so as to absolve the rest, each openly or secretly believing the answer is known already. Some families are so absorbed in the ethical aspects of an ideal image of themselves that they literally cannot believe a therapist could be interested in anything else. A whole child-rearing mythology has arisen, attributed to the discoveries of Freud, that "You always blame the parents." Further normative assumptions about the place of a child in the family, the duties of parents, the value of "closeness" between family members, or of authority, obedience, or "communication" are usually so taken for granted in terms of the parents own upbringing (and they tend to pick marital partners with rather similar assumptions based upon social class, ethnicity, education, religious origin, etc.) that they are seldom examined at all, much less considered in an historical dimension.

Yet the historical perspective can illuminate human conflicts that are far more than personal, assumptions that are still but dimly examined, certainly far from definitively analyzed. Tocqueville was struck by the loosened bonds of authority in the American family—for example, by the heightened intimacy between fathers and sons that was achieved as strict obedience was less valued— but was hard put to explain what happened to the American girl who grew into a married woman. He saw the independent, surprisingly mature, clear-thinking young unmarried woman surrender herself voluntarily to the "yoke" of marriage and thereafter become less active beyond her own hearth than the European woman with whom he was more familiar, though enjoying a peculiar respect from her husband while she engaged in a courageous struggle to fulfill the family role to which she was assigned. Lacking the excuse of parental pressure for an arranged marriage, the

American female virtuously, if perhaps a trifle coldly (Tocqueville felt), completed the tasks she and her husband expected of her. Included as a preeminent responsibility was that of raising children. The woman had also the "moral" leadership of the family, and she was its representative in ensuring a continuity at least of Christian religious customs. But the overall position and life course of the American woman seems to have become, with the passage of time, almost more puzzling than ever.

Linked to the expectations of women that Tocqueville saw as a consequence of political cum social democracy and a Puritan heritage has been a changing view of childhood. Whether we accept Ariès's view that childhood as a special condition or period was a possibly unfortunate invention of modernists or DeMause's idea that the treatment of children is gradually improving over the ages, the development of American family self-perceptions has depended in very substantial degree on American notions of *what children need*, and *what a mother should give children*, and *what children could expect from a father*.

As Dr. Davis so aptly reminds us, every parental generation has some belief in a golden age of family stability. As like as not, though, the belief has been expressed as a perpetual mourning in every historical period about the way children turn out. Young people, in *our* day, were not the same. Meaning, we weren't. Our own parents, it is said, were more respected, if not feared. Of course, we may be *closer* to our children (smugly); we may have treated our children as innocent creatures rather than repositories of original sin; we may see them as inhabited by infant sexuality rather than evil tendencies toward masturbation, or as needful of physical activity rather than swaddling; but it seems particularly cruel and undeserved if this kindness and understanding is repaid by the drop-out delinquent who enters one drug culture after another or the tense uneasy teenager who contemplates suicide. And authority once yielded, it seems, can never be regained. Our own innocence cannot be reclaimed—once you have a relationship with a child-person that abjures hitting, you cannot, without damage to either of you, start to hit. If fresh means of achieving psychological distance in the midst of closeness have been adopted, they will be hard to give up. A sort of privacy has joined our values, perhaps in part as an increased housing affluence made that possible. It is forgotten that the 19th-century golden age supported a parental intrusiveness that the mid-20th century will not tolerate in the midst of its family closeness.

This was especially a value in the upbringing of daughters, who were expected not to have any secrets from their mothers; yet boys were subject to special humiliations. Sons were expected to bend over and expose their buttocks for a whipping if necessary, and the fatherly right to demand this subjugation in most cases disappeared only when the son became big enough to "whip" the father. I have interviewed a fair number of men, now parents themselves, who describe rather vividly some crucial instance where the father attempted to beat his son and was himself knocked down or at least refused obedience. What interested me most in these accounts was that I never encountered one case where (so far as the younger men knew) the defeated father held the son's rebellion against him. One could surmise that a similar incident may have occurred in the father's own emancipation.

But emancipations occur differently now, and we are not sure if we like the change.

Will our anxiety lead to a cyclical return, in some degree, to ideals of parent-child relationships that we seek to recover, however difficult that recovery may be? If so, the effect on clinical work with families could be enormous. But another development seems likely to take precedence.

One change in outlook that seems primarily of our own century—the past 75 years or so—can be expected to lessen in certain areas. That is our emphasis on the importance of *relationships* in the family and of physical nearness as a necessity for developing these relations.

This emphasis seems to me an attitude especially American, though 20th-century psychological writings have emphasized worldwide the effects of childhood experiences in the family. It is particularly in America, for example, that mothers feel most crucially a conflict about being physically away from their children, whether through employment or other activity. Women in cultures that depend more on the extended family have never accepted, so much as the conflicted American women, that there is some special importance to a physical presence in the home or a touchingly personal child-care. English upper-middle-class mothers who felt happy about sending their children to boarding school early have been less troubled about following their own careers. Where most well-to-do families in a culture have nannies or ayahs to care for the children, it has meant that both women servants and employers had less to do with their children. Kibbutz ideals, in an Israeli society otherwise rather ideologically conservative, place little accent on hours of togetherness. The Soviet culture puts great emphasis on children but not upon mothers or fathers being present.

Black society in the United States, whether from economic necessity or as a resource developed during slavery, or perhaps related to certain African heritages, has tended to bring into child-care activities a variety of women kin who are treated not as poor substitutes for a "real" mother but as mothering persons in their own right. As Dr. Davis indicates, however, when a golden age of family stability in America is seen as a glorious togetherness of all generations, this ignores not only the disruptions of slavery but the earlier death of grandparents in former times; nor is the tradition of emigrating, moving westward, or going North generally considered. Many young people in earlier American days said good-bye not only to grandparents but to their parents, often forever. Most child care in this country has tended to fall on the mother of a nuclear family. Population was not sufficiently plentiful nor money so much available that wet nurses could be employed. The conception of earlier European centuries that one made arrangements for the proper care of children (alas, sometimes not too proper for the child's health and safety) but saw no special virtue in performing the tasks *oneself* has largely died out in the American mother's perceptions.

The medieval-into-modern European ideal of parent care was that the child should be *provided* for: fed, given a place to sleep, clothes, educated for his proper station in life, trained. There was no repugnance to early apprenticeships or sending children into domestic service or factory labor, since these helped to provide for the child and train him in a way of caring for his own needs. Nor, apparently, was there a great difference in the way mothers and fathers were seen in this respect. The conception of a woman as an especially "tender" person whose personal touch with her child was essential seems to have developed through the 19th century and achieved its fullest expression in the 20th century, with its psychology of infant development in the hands of a desirable mother.

This attitude is now being challenged by a societal alteration: the movement toward an emancipation of women in terms of employment. Meanwhile, however, the philosophical stream of a child's requiring the "tender" immediate care of a mother has by no means worn itself out; yet the equality of women in the professional marketplace has still not been achieved. At the same time, a woman not gainfully employed is seen increasingly as both foolish and self-indulgent by almost everybody but Phyllis Schafly. The divorce rate remains high, with child custody going in over 90% of cases to the mother, but alimony beyond minimal "child support" has gone out of fashion except among the very rich. In other words, the present-day American woman has assumed a cluster of family responsibilities to a degree almost unprecedented in recorded history. Women are expected both to cuddle their children and support them, often without the aid of a father and without the power or the psychological rewards of a true matriarchy.

It is in this role that the next clarifications and changes may be expected, though the directions may not be so clear. Even alternative life styles and commune set-ups still lean toward a mother's physical presence with children, rather than the father's. In Dr. Eiduson's research project in California, it was found that commune fathers were present only 25% of the time when interviewers went to visit a family during the day; it was the mother who expected to be interviewed.[1] Efforts toward shorter work weeks and workdays may yield more division of parental care between mother and father but perhaps fewer times of their both being present to develop a relationship with each other.

Professor Davis made the point of an increased moral sensibility as related to family life over the past two centuries. The current public uproar about child pornography and wife beating, with its acknowledgement of a continuance of exploitation into the present

---

[1]See Chapter 5 for further details of this study.

day, probably means that we can tolerate knowning more about abuses as we feel more capable of rooting them out. However, in avoiding the fallacy of idealizing the past, we should also be most chary of gilding the present.

Economic conditions have improved enourmously, but payments for welfare mothers and their children are talked about, very often, in much the same terms as 18th-century gentlemen and ladies proposed workhouses for the poor.

Further shifts can be anticipated in regard to whom in the family we feel most moral *about*: Are we to concentrate on exterminating the measure of child abuse that still exists, or on becoming more sensitive to the plight of dilemma-ridden mothers? Or is it the father's alienation from his children (divorced or not) that will receive most attention? Who will be considered an essential family—one parent and child, two parents and child, a man and a woman without children? The existence of a marriage document seems less and less to be considered important unless there are children expected, while at the same time childless unions become more acceptable. Various group-living arrangements have been attempted but so far have not really "caught on" after a brief vogue among the less conventional.

An individual who seeks help from the professional clinician tends to see his or her trouble in almost purely personal terms, even if the world is castigated. It seems really not to dawn on the person in trouble that a difficulty is *not* simply unique, that he or she may not be at fault if someone else in the family isn't to blame. The clinician can help both individuals and families to lift their perspectives out of what may be realistically unique, to perceive as well the conflicts that are inherent in family relationships, dilemmas that are seldom entirely new and that have been handled in a variety of ways at different periods. Meanwhile the family continues to survive, and perhaps to move under our very explorations into a *terra incognita* that we have yet to recognize.

II

Stratification: Class, Race, and the Family

2

The Effects of Social Class on Parental Values and Practices

Melvin L. Kohn

My thesis is straightforward and relatively simple: that there are substantial differences in how parents of differing social-class position raise their children, that these differences in parental practices result chiefly from class differences in parents' values for their children; and that such class differences in parental values result in large measure from differences in the conditions of life experienced by parents at different social-class levels. This essay attempts to spell out this thesis more concretely and explicitly.[1] Without get-

---

[1] This essay does not purport to be a systematic review of the entire research literature on social class and parent-child relationships. Instead, it focuses on my own research and that of my NIMH colleagues, as well as that of other investigators who have dealt with the same research issues. For a definitive review of the literature on social class and parent-child relationships through the mid-1950s, see Bronfenbrenner (1958). Research of the next decade or so is summarized in Kohn (1969). A seminal essay that reviews and assesses more recent work in the field, and also appraises the major interpretations of social class and parent-child relationships, is that by Gecas (1977).

Melvin L. Kohn • National Institute of Mental Health, Bethesda, Maryland.

ting into technical aspects of methodology, it also attempts to give some idea of the type of empirical evidence on which the generalizations are based.

SOCIAL CLASS

Since the heart of the thesis is that parents' social-class positions profoundly affect their values and child-rearing practices, it is well to begin by defining *social class*. I conceive of social classes as aggregates of individuals who occupy broadly similar positions in a hierarchy of power, privilege, and prestige.[2] The two principal components of social class, according to most empirical evidence, are education and occupational position. Contrary to the impression of most laymen, income is of distinctly secondary importance, and subjective class identification is virtually irrelevant. The stratificational system of the contemporary United States is probably most accurately portrayed as a continuum of social class positions—a hierarchy, with no sharp demarcations anywhere along the line.[3] For convenience, though, most research on social class and parent-child relationships employs a somewhat oversimplified model, which conceives of American society as divided into four relatively discrete classes: a small "lower class" of unskilled manual workers, a much larger "working class" of manual workers in semiskilled and skilled occupations, a large "middle class" of white-collar workers and professionals, and a small "elite," differentiated from the middle class not so much in terms of occupation as of wealth and lineage. The middle class can be thought of as comprising two distinguishable segments: an upper-middle class of professionals, proprietors, and managers, who generally have at least some college training; and a lower-middle class of small shopkeepers, clerks, and salespersons, generally with less education.

It is probably unnecessary to underline education's importance for placing people in the social order, and it is self-evident that level of educational attainment can be treated as a quantitative variable: a college graduate unequivocally has higher educational credentials than does a high school dropout. But it may be less

---

[2]See Williams (1960, p. 98). See also Barber (1968, p. 292).
[3]See Kohn (1969, pp. 129-131). See also Kohn (1977).

apparent that occupational position is also a major criterion of ranking in this—and in all other—industrial societies. One of the most important and general findings in social science research is the relative invariance of people's ratings of occupational prestige, regardless of which country is studied. This finding is of great theoretical importance in its implication that the stratification system is much the same across all industrialized societies.

As a methodological aside, I want to note that our knowledge of the stability of occupational prestige rankings is the result of a long series of studies by many investigators. The first major work in this area was a 1946 cross-sectional survey of the U.S. population by the National Opinion Research Center.[4] At that time, the American population agreed to a remarkably high degree on the relative prestige of various occupations: regardless of which segment of the population was examined, and regardless of people's own occupational levels, most Americans ranked occupations similarly, in a regular and nearly invariant hierarchy from bootblack to physician. Later studies showed that this pattern remained stable over the next quarter-century and that it applied to various special subpopulations, even to children as young as 9 years of age.[5] Another major step in the process of discovery was a 1956 reanalysis of studies of occupational prestige in six industrial societies: the United States, Great Britain, Japan, New Zealand, the Union of Soviet Socialist Republics, and the German Federal Republic.[6] Extremely high intercorrelations (mainly in the .90s) were discovered among these countries despite their cultural differences and despite the inclusion in the analysis of a major noncapitalist state. The stratification system thus appears to be much the same in all industrialized societies. More recent studies have extended this finding to many other countries, several of them non-Western, several of them noncapitalist, and some of them nonindustrialized or only partially so.[7] The evidence for the universality or near-

---

[4]See Hatt (1950), North and Hatt (1953), and Reiss, Duncan, Hatt, and North (1961).
[5]See Gusfield and Schwartz (1963), Hodge, Siegel, and Rossi (1964), and Simmons and Rosenberg (1971).
[6]See Inkeles and Rossi (1956).
[7]See Tiryakian (1958), Svalastoga (1959), Thomas (1962), Hodge, Treiman, and Rossi (1966), and Haller and Lewis (1966).

universality of occupationally based stratification systems is considerable.

These facts are impressive in themselves, and they become even more impressive when we recognize that people's positions in the class system are related to virtually every aspect of their lives: their political party preferences, their sexual behavior, their church membership, even their rates of ill health and death.[8] Among these various phenomena, none, certainly, is more important than the relationship of social class to parental values and child-rearing practices. But it is well for us to be aware, when we focus on this relationship, that it is one instance of a much larger phenomenon: the wide ramifications of social stratification for people's lives. Any interpretation we develop of the relationship between social class and parental values and behavior must be applicable, at least in principle, to the larger phenomenon as well.

Social class has proved to be so useful a concept in social science because it refers to more than simply educational level, or occupation, or any of the large number of correlated variables. It is useful because it captures the reality that the intricate interplay of all these variables creates different basic conditions of life at different levels of the social order. Members of different social classes, by virtue of enjoying (or suffering) different conditions of life, come to see the world differently—to develop different conceptions of social reality, different aspirations and hopes and fears, different conceptions of the desirable.

The last is particularly important for our purposes, because conceptions of the desirable—that is, values—are a key bridge between position in the larger social structure and behavior. Of particular pertinence to our present interests are people's values for their children.

PARENTAL VALUES

By values, I mean standards of desirability—criteria of preference.[9] By parental values, I mean those standards that parents

---

[8]See Berelson and Steiner (1964). See also the references in Kohn (1969, p. 3, note 2).

[9]See Williams (1968, p. 283).

would most like to see embodied in their children's behavior. Since values are hierarchically organized, a central manifestation of value is to be found in choice. For this reason, most studies of parental values require parents to choose, from among a list of generally desirable characteristics, those few that they consider most desirable of all, and, in some studies, those that they consider the least important, even if desirable.[10] Such a procedure makes it possible to place parents' valuations of each characteristic on a quantitative scale. We must recognize that parents are likely to accord high priority to those values that are not only important, in that failing to achieve them would affect the children's futures adversely, but also problematic, in that they are difficult of achievement. Thus, the indices of parental values used in most of the pertinent inquiries measure conceptions of the "important, but problematic."[11]

There have been two central findings from these studies. One is that parents at all social-class levels value their children's being honest, happy, considerate, obedient, and dependable.[12] Middle- and working-class parents share values that emphasize, in addition to children's happiness, their acting in a way that shows respect for the rights of others. All class differences in parental values are variations on this common theme.

Nevertheless, there are distinct differences in emphasis between middle- and working-class parents' values. The higher a parent's social-class position, the more likely he is to value characteristics indicative of self-direction and the less likely he is to value characteristics indicative of conformity to external authority.[13] That is, the higher a parent's social-class position, the greater the likelihood that he will value for his children such characteristics as consideration, an interest in how and why things happen, responsibility, and self-control, and the less the likelihood that he will value such characteristics as manners, neatness and cleanliness, being a good student, honesty, and obedience. More detailed analyses show that the differential evaluation of self-direction and conformity to external authority by parents of varying social-class posi-

---

[10]See Kohn (1969, pp. 18-19, 47-48).
[11]See Kohn (1969, pp. 23-24).
[12]See Kohn (1969, pp. 20-21, 42-43, 50-51).
[13]See Kohn (1969, Chapters 2, 3, 4; 1977).

tion obtains whatever the age and sex of the child, in families of varying size, composition, and functional pattern.[14]

This essential finding has been repeatedly confirmed, both for fathers and for mothers. The original finding came from a small study in Washington, D.C., in the late 1950s, but it has since been confirmed in several other U.S. studies, including three nation-wide studies, one as recent as 1975. It has also been confirmed in studies in Italy, Germany, Great Britain, France, Ireland, and Taiwan. There are no known exceptions.[15]

The correlations of class with parental valuation of the individual characteristics (e.g., self-control and obedience) are not very large: none is larger than .20. The correlation of class with an overall index of valuation of self-direction or conformity, based on factor analysis, is a more substantial .34. But even a correlation of .34 is, by absolute standards, only moderate. What makes the class differences in parental values impressive is their consistency. It has repeatedly been confirmed that social class continues to be nearly as strongly correlated with parents' valuation of self-direction when all other major lines of social demarcation—national background, religious background, urbanicity, region of the country, and even race—are statistically controlled.[16] Social class, in fact, is as strongly correlated with parental valuation of self-direction as are all these other major lines of social demarcation combined.[17] Thus, social class, even though only moderately correlated with parental values, stands out as the single most important social influence on parents' values for their children.

Parenthetically, it is pertinent to ask whether there have been changes in parental values, especially in parental valuation of

---

[14]See Kohn (1969, pp. 48-59, 68-69).

[15]See Kohn (1969, Chapters 2, 3, 4), Olsen (1971), Perron (1971), Platt (n.d.), Hynes (1977), Hoff and Grueneisen (1977a,b), Bertram (1976, 1977), Franklin and Scott (1970), Clausen (1974), Campbell (1977), Wright and Wright (1976), and Kohn (1976d). See also Le Masters (1975) and Sennett and Cobb (1973).

[16]See Kohn (1969, pp. 59-72), Wright and Wright (1976), Kohn (1976d), Kohn (1977).

[17]This is not to say that other major lines of social demarcation are unimportant for parental values. Race, for example, has a consistent effect on parental values at all social-class levels. But the effect of race on parental values is only about one-fourth as great as that of social class.

self-direction or conformity to external authority, over the few years for which data are available. The evidence, unfortunately, is equivocal. There is no substantial evidence that there have been changes, but also no conclusive evidence that there have not. In any case, the magnitude of the correlation between social class and parental valuation of self-direction is as strong in the latest available data, a national survey conducted in 1975, as in earlier studies.[18] The class-values relationship is as important for understanding parental values in the mid-1970s as it was in the mid-1950s.

PARENTAL VALUES AND PARENTAL PRACTICES

We would have little interest in parental values but for our belief that parents' values affect their child-rearing practices. The evidence here is much less definitive than on the relationship of class to parental values, but what evidence we do have is altogether consistent. Parents do behave in accord with their values in the two important realms where the question has been studied: in their disciplinary practices and in the allocation of parental responsibilities for imposing constraints on, and providing emotional support for, their children.

Disciplinary Practices

Most early research on class differences in disciplinary practices was directed toward learning whether working-class parents typically employ techniques of punishment different from those used by the middle class. In his definitive review of the research literature on social class and family relationships through the mid-1950s, Bronfenbrenner (1958, p. 424) summarized the results of the several relevant studies as indicating that "working-class parents are consistently more likely to employ physical punishment, while middle-class families rely more on reasoning, isolation, appeals to guilt, and other methods involving the threat

---

[18]This statement is based on my unpublished analysis of the data of the 1975 National Opinion Research Center's General Social Survey as compared to earlier studies cited above.

of loss of love." This conclusion has been challenged in later research.[19] Whether or not it is still true, the difference in middle-
and working-class parents' propensity to resort to physical
punishment certainly never has been great.

For our purposes, in any case, the crucial question is not which
disciplinary method parents prefer but when and why they use
one or another method of discipline. The early research tells us
little about the when and why of discipline; most investigators had
relied on parents' generalized statements about their usual or their
preferred methods of dealing with disciplinary problems, irrespec-
tive of what the particular problem might be. But surely not all
disciplinary problems evoke the same kind of parental response. In
some sense, after all, the punishment fits the crime. Under what
conditions do parents of a given social class punish their children
physically, reason with them, isolate them—or ignore their actions
altogether?

Recent studies have shown that neither middle- nor working-
class parents resort to punishment as a first recourse when their
children misbehave.[20] It seems instead that parents of both social
classes initially post limits for their children. But when children
persist in misbehavior, parents are likely to resort to one or another
form of coercion. This is true of all social-class levels. The principal
difference between the classes is in the specific conditions under
which parents—particularly mothers—punish children's misbehavior. Working-class parents are more likely to punish or refrain from punishing on the basis of the direct and immediate consequences of children's actions, middle-class parents on the basis
of their interpretation of children's intent in acting as they do.[21]
Thus, for example, working-class parents are more likely to punish
children for fighting than for arguing with their brothers and sis-

---

[19]See Erlanger (1974). Erlanger found that subsequent studies failed to confirm the
pattern Bronfenbrenner had discovered in the earlier studies, particularly as regards working-class parents' allegedly greater propensity to use physical punishment.

[20]See Kohn (1969, pp. 92-95, 102-103).

[21]See Kohn (1969, Chapter 6) and Gecas and Nye (1974). Class differences in the
conditions under which parents publish their children's misbehavior probably do
not begin until the children are about 6 years old, but they seem to apply regardless of age thereafter, at least until the mid-teens.

ters and are also more likely to punish for aggressively wild play than for boisterousness—the transgression in both instances being measured in terms of how far the overt action transgresses the rules. Middle-class parents make no such distinction. But they do distinguish, for example, between wild play and a loss of temper, tolerating even excessive manifestations of the former as a childish form of emotional expression, but punishing the latter because it signifies a loss of mastery over self.

To say that working-class parents respond more to the consequences of children's misbehavior and middle-class parents more to their own interpretation of the children's intent gets dangerously close to implying that while middle-class parents act on the basis of long-range goals for children, working-class parents do not. On the contrary, the evidence suggests that parents of both social classes act on the basis of long-range goals—but that the goals are different. The interpretive key is provided by our knowledge of class differences in parental values. Because middle- and working-class parents differ in their values, they view children's misbehavior differently; what is intolerable to parents in one social class can be taken in stride by parents in the other. In both social classes, parents punish children for transgressing important values, but since the values are different, the transgressions are differently defined. If self-direction is valued, transgressions must be judged in terms of the reasons why the children misbehave. If conformity to external authority is valued, transgressions must be judged in terms of whether or not the actions violate externally imposed proscriptions.

The Allocation of Parental Responsibilities for Support and Constraint

The connection between values and punishment of disvalued behavior is direct: punishment is invoked when values are transgressed. There are also less direct but broader behavioral consequences of class differences in parental values. In particular, class differences in parental values have important consequences for the overall patterning of parent-child interaction.

In common with most investigators, I conceive of parent-child relationships as structured along two principal axes: support and constraint. This conception is derived in part from Parsons and Bales's (1955) theoretical analysis of family structure and in part from Schaefer's (1959) empirical demonstration that the findings of several past studies of parent-child relationships could be greatly clarified by arraying them along these two dimensions.

Because their values are different, middle- and working-class parents evaluate differently the relative importance of support and constraint in child rearing. One would expect middle-class parents to feel a greater obligation to be supportive, if only because of their concern about children's internal dynamics. Working-class parents, because of their higher valuation of conformity to external rules, should put greater emphasis upon the obligation to impose constraints. We should therefore expect the ratio of support to constraint in parents' handling of their children to be higher in middle-class than in working-class families. And this, according to Bronfenbrenner (1958, p. 425), is precisely what has been shown in those studies that have dealt with the overall relationship of parents to child:

> Parent-child relationships in the middle class are consistently reported as more acceptant and equalitarian, while those in the working class are oriented toward maintaining order and obedience. Within this context, the middle class has shown a shift away from emotional control toward freer expression of affection and greater tolerance of the child's impulses and desires.

Whatever relative weight parents give to support and constraint, the process of child rearing requires both. These responsibilities can, however, be apportioned between mother and father in any of several ways. Mothers can specialize in providing support, fathers in imposing constraints; both parents can play both roles more-or-less equally; mothers can monopolize both roles, with fathers playing little part in child rearing; and there are other possible, but less likely, arrangements. Given their high valuation of self-direction, middle-class parents—mothers and fathers both—should want fathers to play an important part in providing support to the children. It would seem more appropriate to working-class parents' high valuation of conformity to external authority that fathers' obligations should center on the imposition of constraints.

The pertinent studies show that in both the middle class and the working class, mothers would prefer to have their husbands play a role that facilitates children's development of valued characteristics.[22] To middle-class mothers, it is important that children be able to decide for themselves how to act and that they have the personal resources to act on these decisions. In this conception, fathers' responsibility for imposing constraints is secondary to their responsibility for being supportive; in the minds of some middle-class mothers, for fathers to take a major part in imposing constraints interferes with their ability to be supportive. To working-class mothers, on the other hand, it is more important that children conform to externally imposed rules. In this conception, the fathers' primary responsibility is to guide and direct the children. Constraint is accorded far greater value than it has for the middle class.

Most middle-class fathers seem to share their wives' views of fathers' responsibilities toward sons and act accordingly.[23] They accept less responsibility for being supportive of daughters—apparently feeling that this is more properly the mothers' role. But many working-class fathers do not accept the obligations their wives would have them assume, either toward sons or toward daughters.[24] These men do not see the constraining role as any less important than their wives do, but many of them see no reason why fathers should have to shoulder this responsibility. From their point of view, the important thing is that children be taught what limits they must not transgress. It does not particularly matter who does the teaching, and since mothers have primary responsibility for child care, the job should be theirs. Of course, there will be occasions when fathers have to backstop their wives. But there is no ideological imperative that makes it the fathers' responsibility to assume an important part in child rearing. As a consequence, many working-class fathers play little role in child rearing, considering it to be their wives' proper responsibility.[25]

---

[22]See Kohn (1969, pp. 120-122).

[23]See Kohn (1969, pp. 113-114).

[24]See Kohn (1969, pp. 120-122).

[25]The general picture of class differences in parental role-allocation sketched above appears to apply regardless of the age of the child, beginning at least as young as age 10 or 11 and probably earlier.

Theories of personality development, including Parsons and Bales's (1955) sociological reinterpretation of the classical Freudian developmental sequence, have generally been based on the model of a family in which the mothers' and fathers' intrafamily roles are necessarily differentiated, with mothers specializing in support and fathers in constraint.[26] However useful a first approximation this may be, both middle- and working-class variations on this general theme are sufficiently great to compel a more precise formulation.

The empirical evidence is partly consistent with the mother-supportive, father-constraining formulation, for even in middle-class families, almost no one reports that fathers are more supportive than mothers. Yet, in a sizable proportion of middle-class families, mothers take primary responsibility for imposing constraints on sons, and fathers are at least as supportive as mothers. And although middle-class fathers are not likely to be as supportive of daughters as their wives are, it cannot be said that fathers typically specialize in constraint, even with daughters.

It would be a gross exaggeration to say that middle-class fathers have abandoned the prerogatives and responsibilities of authority in favor of being friends and confidants to their sons. Yet the historical drift is probably from primary emphasis on imposing constraints to primary emphasis on support.[27] In any event, mothers' and fathers' roles are not sharply differentiated in most middle-class families; both parents tend to be supportive. Such division of functions as exists is chiefly a matter of each parent's taking special responsibility for being supportive of the children of the parent's own sex.

Mothers' and fathers' roles are more sharply differentiated in working-class families, with mothers almost always being the more supportive. Yet, despite the high valuation put on the constraining function, fathers do not necessarily specialize in setting limits, even for sons. In some working-class families, mothers specialize in support, fathers in constraint; in many others, the division of responsibilities is for the mothers to raise the children, the fathers

---

[26]See also Bronfenbrenner (1960) and Kohn (1969, p. 124, note 11).
[27]See Bronson, Katten, and Livson (1959) and Bronfenbrenner (1961).

to provide the wherewithal. This pattern of role allocation probably is and has been far more prevalent in American society than the formal theories of personality development have recognized.

SOCIAL CLASS, VALUES, AND CONDITIONS OF LIFE

There are, then, remarkably consistent relationships between social class and parental values and behavior. But we have not yet touched on the question: Why do these relationships exist? In analytic terms, the task is to discover which of the many conditions of life associated with class position are most pertinent for explaining why class is related to parental values. Since many of the relevant conditions are implicated in people's occupational lives, our further discussion is focused on one crucial set of occupational conditions: those that determine how much opportunity people have to exercise self-direction in their work.

The principal hypothesis that has guided this line of research is that class-correlated differences in people's opportunities to exercise occupational self-direction—that is, to use initiative, thought, and independent judgment in work—are basic to class differences in parental values. Few other conditions of life are so closely bound up with social class position as are those that determine how much opportunity, even necessity, people have for exercising self-direction in their work. Moreover, there is an appealing simplicity to the supposition that the experience of self-direction in so central a realm of life as work is conducive to valuing self-direction, off as well as on the job, and to seeing the possibilities for self-direction not only in work but also in other realms of life.

Although many conditions of work are either conducive to or deterrent of the exercise of occupational self-direction, three in particular are critical.

First, a limiting condition: people cannot exercise occupational self-direction if they are closely supervised. Not being closely supervised, however, does not necessarily mean that people are required—or even free—to use initiative, thought, and independent judgment; it depends on how complex and demanding is their work.

A second condition for occupational self-direction is that work allow a variety of approaches; otherwise the possibilities for exercising initiative, thought, and judgment are seriously limited. The organization of work must not be routinized; it must involve a variety of tasks that are in themselves complexly structured.

The third and most important determinant of occupational self-direction is that work be substantively complex. By the *substantive complexity* of work I mean, essentially, the degree to which performance of that work requires thought and independent judgment. All work involves dealing with things, with data, or with people; some jobs involve all three, others only one or two of these activities. Work with things can vary in complexity from ditch-digging to sculpturing; similarly, work with people can vary in complexity from receiving simple instructions to giving legal advice; and work with data can vary from reading instructions to synthesizing abstract conceptual systems. Although, in general, work with data or with people is likely to be more complex than work with things, this is not always the case, and an index of the overall complexity of work should reflect its degree of complexity in each of these three types of activity. What is important about work is not whether it deals with things, with data, or with people, but its complexity.

No one of these occupational conditions—freedom from close supervision, nonroutinization, and substantive complexity—is definitional of occupational self-direction. Nevertheless, each of these three conditions tend to be conducive to the exercise of occupational self-direction, and the combination of the three both enables and requires it. Insofar as people are free of close supervision, work at nonroutinized tasks, and do substantively complex work, their work is necessarily self-directed. And insofar as they are subject to close supervision, work at routinized tasks, and do work of little substantive complexity, their work does not permit self-direction.

THE RELATIONSHIP OF OCCUPATIONAL SELF-DIRECTION TO PARENTAL VALUES

Since most of the research on the relationship between occupational self-direction and parental values deals only with men's occupational conditions and men's values, I shall first discuss fathers' values, then broaden the discussion to include mothers' values as well. All three occupational conditions that are determinative of occupational self-direction prove to be empirically related to fathers' values.[28] Men who are free from close supervision, who work at nonroutinized tasks, and who do substantively complex work tend to value self-direction rather than conformity to external authority for their children. This being the case, it becomes pertinent to ask whether the relationship between social class and fathers' values can be explained as resulting from class differences in the conditions that make for occupational self-direction.

It must be emphasized that in dealing with these occupational conditions, we are concerned not with distinctions that cut across social class but with experiences constitutive of class. The objective is to learn whether these constitutive experiences are pertinent for explaining the class relationship. To achieve this objective, we statistically control occupational dimensions that have proved to be related to values and orientation, to determine whether this reduces the correlation between class and fathers' valuation of self-direction or conformity for their children. This procedure is altogether hypothetical, for it imagines an unreal social situation: social classes that did not differ from one another in the occupational conditions experienced by their members. But it is analytically appropriate to use such hypothetical procedures, for it helps us differentiate those occupational conditions that are pertinent for explaining the relationship of class to parental values from those occupational conditions that are not pertinent. In fact, statistically controlling the conditions that make for occupational self-direction reduces the correlation of class to fathers' valuation of self-direction or conformity by nearly two thirds.[29] The lion's share of the reduction is attributable to the substantive complexity of the work, but closeness of supervision and routinization are relevant

[28]See Kohn (1969, Chapter 9), based on data from the United States and Italy. The generalization is also supported by Hoff and Grueneisen (1977a,b), using West German data; by Hynes (1977), using data from Ireland; by St. Peter (1975) using U.S. data; and by Scurrah and Montalvo (1975), using data from Peru; but not by Olsen (1971), using data from Taiwan.
[29]See Kohn (1969, pp. 161-163).

too. By contrast, though, statistically controlling numerous other occupational conditions has a much weaker effect—reducing the class correlation by only one-third.[30] And controlling both sets of occupational conditions reduces the correlation of class to fathers' values by no more than does controlling occupational self-direction alone. Thus, other occupational conditions add little to the explanatory power of the three that are determinative of occupational self-direction.

These findings come mainly from a cross-national study in the United States.[31] They are confirmed by a smaller-scale study in Turin, Italy.[32] A study in Taiwan failed to confirm these findings, but it is impossible to say whether this is because occupational conditions have different consequences in that partially industrialized, non-Western society, or because methodological problems of the Taiwanese study may have obscured the phenomenon.[33] A number of other pertinent but not entirely comparable studies—in Peru, West Germany, Ireland, and the United States— tend to confirm the original U.S. and Italian findings.[34] But definitive confirmation awaits the completion of studies now in progress in Ireland, Poland, and West Berlin.

Much less is known about the relationship between occupational conditions and mothers' values. My colleagues and I are currently analyzing data on the relationship between employed mothers' occupational conditions and their values for their children.[35] Preliminary results indicate that women's occupational conditions affect their values in much the same way as do men's occupational conditions. We are also investigating the relationship between housework, conceptualized in essentially the same way as any other work, and parental values, but on this issue we do not

---

[30]See Kohn (1969, pp. 182-183 and Table 10-7).

[31]See Kohn (1969, pp. 152-164, 182-183).

[32]See Kohn (1969, pp. 143-152).

[33]See Olsen (1971) and Kohn (1977).

[34]See Scurrah and Montalvo (1975), Hoff and Grueneisen (1977a,b), Hynes (1977), and St. Peter (1975).

[35]In 1974, we conducted a 10-year follow-up to the 1964 study of employed men that is the basis of many of the findings presented in Kohn (1969). In the follow-up study, we not only reinterviewed the men, but this time we also interviewed their wives and children (see Kohn, 1977).

yet have findings. There is information from a study in Turin, Italy, that men's occupational conditions affect their wives' values; this appears to be particularly the case in the middle class.[36] But these data say nothing about the mechanisms by which men's occupational conditions affect their wives' values—it may be that men's occupational conditions affect their own values and that men influence their wives; it may be that men communicate something of their occupational experience to their wives and that this knowledge affects the wives' value choices; or it may be any of several other possibilities. These questions, too, are currently being studied in research that attempts to unravel the processes by which each spouse's occupational conditions affect the other's values. Clearly, much remains to be learned. But even now, there is every reason to believe that women's values are affected by class-associated conditions of life through processes similar to those operating for men.

Because the relationship between exercising self-direction on the job and valuing self-direction for children is so direct, one might conclude that parents are simply preparing their children for the occupational life to come. I believe, rather, that parents come to value self-direction or conformity as virtues in their own right, not simply as means to occupational goals. One important piece of evidence buttresses this impression: studies in both the United States and Italy show that the relationship between men's occupational experiences and their values is the same for daughters as for sons, yet it is hardly likely (especially in Italy) that most fathers think their daughters will have occupational careers comparable to those of their sons. It would thus seem that occupational experience helps structure parents' views not only of the occupational world but of social reality in general.

THE DIRECTION OF CAUSAL EFFECTS

It could be argued that the empirical interrelationships of social class, occupational self-direction, and parental values reflect the propensity of people who value self-direction to seek out jobs

---

[36]See Kohn (1969, Chapter 9).

that offer them an opportunity to be self-directed in their work and, once in a job, to maximize whatever opportunities the job allows for exercising self-direction. But we know that occupational choice is limited by educational qualifications, which in turn are greatly affected by the accidents of family background, economic circumstances, and available social resources. Moreover, the opportunity to exercise self-direction in one's work is circumscribed by job requirements. Thus, an executive must do complex work with data or with people; he cannot be closely supervised; and his tasks are too diverse to be routinized—to be an executive requires some substantial self-direction. Correspondingly, to be a semi-skilled factory worker precludes much self-direction. The substance of one's work cannot be especially complex; one cannot escape some measure of supervision; and if one's job is to fit into the flow of other people's work, it must necessarily be routinized. The relationship between being self-directed in one's work and holding self-directed values would thus seem to result not just from self-directed people's acting according to their values but also from job experiences affecting these very values.

This, of course, is an *a priori* argument. But we also have empirical evidence that the most important of the three occupational conditions determinative of occupational self-direction—the substantive complexity of the work—actually does have a causal impact on parental values.[37] This evidence is based on a statistical technique called *two-stage least-squares*, which was developed by econometricians for analyzing reciprocal effects. With this technique, it has been shown that the substantive complexity of work has a causal effect not only on parental values but on people's values and orientation generally, even on their intellectual functioning. These effects are independent of the selection processes that draw men into particular fields of work and independent of men's efforts to mold their jobs to fit their needs and values. Admittedly, cross-sectional data cannot provide definitive evidence of causality—only analyses of longitudinal data measuring real change in real people can be definitive. Nevertheless, these findings do establish a strong *prima facie* case that the substantive com-

---

[37]See Kohn and Schooler (1973).

plexity of work has a real and meaningful effect on parental values and also on a very wide range of psychological processes.

More definitive, albeit less extensive, evidence comes from longitudinal analyses currently in process.[38] Analyses of longitudinal data are immensely difficult because they require the development of "measurement models" that separate unreliability of measurement from real change in the phenomena studied. Such measurement models have thus far been constructed for substantive complexity and for one facet of psychological functioning: intellectual flexibility. The latter was chosen precisely because it appeared to offer the toughest test: intellectual flexibility is obviously pertinent to job placement, and it might be expected to be one of the most stable psychological phenomena.

Stable it certainly is. The correlation between the men's intellectual flexibility at the time they were originally studied and their intellectual flexibility 10 years later, shorn of measurement error, is .93. Nevertheless, the effect of the substantive complexity of work on intellectual flexibility is striking—on the order of one-fourth as great as the effect of the men's earlier levels of intellectual flexibility. Since this analysis is based on men no younger than 26 years of age, who are at least 10 years into their occupational careers, the effect of the substantive complexity of the job on intellectual flexibility is indeed impressive.

The longitudinal analysis demonstrates also something that no cross-sectional analysis could show—that, over time, the relationship between substantive complexity and intellectual flexibility is truly reciprocal. Substantive complexity has a more immediate effect on intellectual flexibility: today's job demands affect today's thinking processes. Intellectual flexibility, by contrast, has a delayed effect on substantive complexity: today's intellectual flexibility has scant effect on today's job demands, but it will have a sizable effect on the further course of one's career. Cross-sectional analyses portray only part of this process, making it seem as if the relationship between the substantive complexity of work and psychological functioning were mainly unidirectional, with work affecting psychological functioning but not the reverse. Longitudi-

---

[38]Kohn and Schooler (1978).

nal analysis portrays a more intricate and more interesting, truly reciprocal, process.

Granted, the research has not yet demonstrated that substantive complexity directly affects parental values. Still, because of its remarkable stability, intellectual flexibility offers the crucial test of the hypothesis that the substantive complexity of work actually has a causal effect on psychological functioning. Moreover, intellectual flexibility is intimately related to valuation of self-direction. It is, in fact, an important link between social class and parents' valuation of self-direction or conformity to external authority.[39] Thus, demonstrating the causal impact of substantive complexity on intellectual flexibility gives us every reason to expect substantive complexity to have a causal impact on parental values too. Further analyses will assess this hypothesized causal impact of substantive complexity—and of other determinants of occupational self-direction—on parental values and also on self-conception and social orientation.

THE ROLE OF EDUCATION IN THE RELATIONSHIP BETWEEN SOCIAL CLASS AND PARENTAL VALUES

Education matters for parental values in part because it is an important determinant of occupational conditions. A major reason for looking to such occupational conditions as substantive complexity, closeness of supervision, and routinization as possible keys to understanding the relationship between social class and parental values is that few other conditions of life are so closely related to educational attainment. This explanation has been confirmed in further analyses that have assessed the effects of education on occupational conditions at each stage of career. Education is a prime determinant, for example, of the substantive complexity of the job; and the substantive complexity of the job, in turn, has an appreciable effect on parental values. It is precisely because education is crucial for the very occupational conditions that most strongly affect parental values that education is so powerfully related to parental values.[40]

---

[39]See Kohn (1969, p. 186).
[40]See Kohn and Schooler (1973).

Education also has important direct effects on parental values, quite apart from its indirect effects mediated through occupational conditions. Education matters, aside from its impact on job conditions, insofar as education provides the intellectual flexibility and breadth of perspective that are essential for self-directed values.[41] Thus education has both direct and indirect effects upon parental values, both types of effect contributing importantly to the overall relationship between social class and parental values.

CONCLUSION

The facts and interpretations reviewed in this paper have many implications for medicine. One set of implications that I have dwelt on at length elsewhere is that these findings may help us interpret the consistent statistical relationship between social class and rates of schizophrenia.[42] They may also help us understand the role of the family in the etiology of schizophrenia. More generally, these findings are pertinent to our conception of what is normal and what is not in family functioning. I have tried in this paper to show that there are considerable variations in normal family functioning and that these variations are to be understood in terms of the actual conditions of life that families encounter. The values and child-rearing practices of American parents must be seen in terms of the realities parents face .

REFERENCES

Barber, B. Social stratification. In D. L. Sills (Ed.), *International Encyclopedia of the Social Sciences*, Vol. 15. New York: Macmillan Company and Free Press, 1968, pp. 288-296.

Berelson, B., and Steiner, G. A. *Human Behavior: An Inventory of Scientific Findings.* New York: Harcourt, Brace and World, 1964.

Bertram, H. Gesellschaftliche und Familiäre Bedingungen Moralischen Urteilens. Unpublished doctoral dissertation, Universität Dusseldorf, 1976.

Bertram, H. Personal communication (unpublished data), 1977.

Bronfenbrenner, U. Socialization and social class through time and space. In E. E. Maccoby, T. M. Newcomb, and E. L. Hartley (Eds.), *Readings in Social Psychology.* New York: Holt, Rinehart and Winston, 1958, pp. 400-425.

[41] Kohn (1969, pp. 186-187).
[42] See Kohn (1973). Pertinent, too, are Kohn (1976a,b).

Bronfenbrenner, U. Freudian theories of identification and their derivatives. *Child Development*, 1960, *31* (March), 15-40.

Bronfenbrenner, U. The changing American child—A speculative analysis. *Journal of Social Issues*, 1961, *17*, 6-18.

Bronson, W. C., Katten, E. S., and Livson, N. Patterns of authority and affection in two generations. *Journal of Abnormal and Social Psychology*, 1959, *58* (March); 143-152.

Campbell, J. D. The child in the sick role: Contributions of age, sex, parental status, and parental values. *Journal of Health and Social Behavior*, 1978, *19* (March), 35-51.

Clausen, J. A. Value transmission and personality resemblance in two generations. Paper presented to the annual meeting of the American Sociological Association, Montreal, August 27, 1974.

Erlanger, H. S. Social class and corporal punishment in childrearing: A reassessment. *American Sociological Review*, 1974, *39* (February), 68-85.

Franklin, J. I. and Scott, J. E. Parental values: An inquiry into occupational setting. *Journal of Marriage and the Family*, 1970, *32* (August), 406-409.

Gecas, V. The influence of social class on socialization. In W. R. Burr, R. Hill, I. L. Reiss, and F. I. Nye (Eds.), *Theories about the Family*. New York: Free Press, 1977.

Gecas, V. and Nye, F. I. Sex and class differences in parent-child interaction: A test of Kohn's hypothesis. *Journal of Marriage and the Family*, 1974, *36* (November); 742-749.

Gusfield, J. R., and Schwartz, M. The meanings of occupational prestige: Reconsideration of the NORC scale. *American Sociological Review*, 1963, *28* (April), 265-271.

Haller, A. O., and Lewis, D. M. The hypothesis of intersocietal similarity in occupational prestige hierarchies. *American Journal of Sociology*, 1966, *72* (September), 210-216.

Hatt, P. K. Occupation and social stratification. *American Journal of Sociology*, 1950, *55* (May), 533-543.

Hodge, R. W., Siegel, P. M., and Rossi, P. H. Occupational prestige in the United States: 1925-1963. *American Journal of Sociology*, 1964, *70* (November), 286-302.

Hodge, R. W., Treiman, D. J., and Rossi, P. H. A comparative study of occupational prestige. In R. Bendix and S. M. Lipset (Eds.), *Class, Status, and Power*, 2nd ed. New York: Free Press, 1966.

Hoff, E.-H. and Grueneisen, V. Arbeitserfahrungen, Erziehungseinstellungen, und Erziehungsverhalten von Eltern. In H. Lukesch and K. Schneewind (Eds.), *Familiäre Sozialisation: Probleme, Ergebnisse, Perspektiven*. Stuttgart: Klett, 1977.(a)

Hoff, E.-H. and Grueneisen, V. Personal communication (unpublished data), 1977. (b)

Hynes, E. Personal communication (unpublished data), 1977.

Inkeles, A., and Rossi, P. H. National comparisons of occupational prestige. *American Journal of Sociology*, 1956, *61* (January), 329-339.

Kohn, M. L. *Class and Conformity: A Study in Values*. Homewood, Ill.: Dorsey, 1969. (Second edition, 1977, published by the University of Chicago Press.)

Kohn, M. L. Social class and schizophrenia: A critical review and a reformulation. *Schizophrenia Bulletin*, 1973, Issue 7 (Winter), 60-79.

Kohn, M. L. The interaction of social class and other factors in the etiology of schizophrenia. *American Journal of Psychiatry,* 1976, *133* (February), 177-180. (a)

Kohn, M. L. Looking back—A 25-year review and appraisal of social problems research. *Social Problems,* 1976, *24* (October), 94-112. (b)

Kohn, M. L. Occupational structure and alienation. *American Journal of Sociology,* 1976, *82* (July), 111-130. (c)

Kohn, M. L. Social class and parental values: Another confirmation of the relationship. *American Sociological Review,* 1976, *41* (June), 538-545. (d)

Kohn, M. L. Reassessment, 1977. In *Class and Conformity: A Study in Values,* 2nd ed. Chicago: University of Chicago Press, 1977, pp. xxv-lx.

Kohn, M. L., and Schooler, C. Occupational experience and psychological functioning: An assessment of reciprocal effects. *American Sociological Review,* 1973 *38* (February), 97-118.

Kohn, M. L., and Schooler, C. The reciprocal effects of the substantive complexity of work and intellectual flexibility: A longitudinal assessment. *American Journal of Sociology,* 1978, *84* (July), 24-52.

Le Masters, E. E. *Blue-Collar Aristocrats: Life-Styles at a Working-Class Tavern.* Madison: University of Wisconsin Press, 1975.

North, C. C., and Hatt, P. K. Jobs and occupations: A popular evaluation. In R. Bendix and S. M. Lipset (Eds.), *Class, Status, and Power.* Glencoe, Ill.: Free Press, 1953, pp. 411-426.

Olsen, S. M. Family, Occupation, and Values in a Chinese Urban Community. Unpublished Ph.D. dissertation, Cornell University, 1971.

Parsons, T., and Bales, R. F. *Family, Socialization and Interaction Process.* Glencoe, Ill.: Free Press, 1955.

Perron, R. *Modèles d'enfants, enfants modèles.* Paris: Presses Universitaires de France, 1971.

Platt, J. Social class and childrearing norms in Britain and the U.S.A. Unpublished manuscript, University of Sussex, n.d.

Reiss, A. J., Duncan, O. D., Hatt, P. K., and North, C. C. *Occupations and Social Status.* New York: Free Press, 1961.

St. Peter, L. G. Fate Conceptions: A Look at the Effects of Occupational Tasks on Human Values. Unpublished doctoral dissertation, University of Nebraska, 1975.

Schaefer, E. S. A circumplex model for maternal behavior. *Journal of Abnormal and Social Psychology,* 1959, *59* (September), 226-235.

Scurrah, M. J., and Montalvo, A. *Clase Social y Valores Sociales en Peru.* Lima, Peru: Escuela de Administracion de Negocios Para Graduados (Serie: Documento de Trabajo No. 8), 1975.

Sennett, R., and Cobb, J. *The Hidden Injuries of Class.* New York: Knopf, 1973.

Simmons, R. G., and Rosenberg, M. Functions of children's perceptions of the stratification system. *American Sociological Review,* 1971, *36* (April), 235-249.

Svalastoga, K. *Prestige, Class and Mobility.* Copenhagen: Glydendal, 1959.

Thomas, R. M. Reinspecting a structural position on occupational prestige. *American Journal of Sociology,* 1962, *67* (March), 561-565.

Tiryakian, E. A. The prestige evaluation of occupations in an underdeveloped country: The Philippines. *American Journal of Sociology,* 1958, *63* (January), 390-399.

Williams, R. M. Jr. American Society: A Sociological Interpretation, 2nd ed. New York: Knopf, 1960.

Williams, R. M. Jr. The concept of values. In D. L. Sills (Ed.), *International Encyclopedia of the Social Sciences*, Vol. 16. New York: Macmillan Company and Free Press, 1968, pp. 283-287.

Wright, J. D., and Wright, S. R. Social class and parental values for children: A partial replication and extension of the Kohn thesis. *American Sociological Review*, 1976, 41 (June), 527-537.

2

Practical Considerations

Norman A. Scotch

It is always a pleasure to read a chapter by Dr. Kohn. One can be sure of impeccable scholarship and considerable thought. Such has been the case with this chapter. Of course, in any one paper, it is impossible to address each and every issue connected with the topic. There are several important points I am sure Dr. Kohn would have raised had he the necessary time. I would like to briefly mention some of these:

1. Dr. Kohn points out that predominant child-rearing processes within social classes differ in important ways from class to class. Another important factor he did not discuss is the variation in practices within classes and their causes and consequences.
2. A major factor in relations between parents and children are factors external to the family. Thus war, economic crises, or perhaps even ideological movements such as the current women's movement often influence the pattern of familial relations, between and within classes, and thus deserve consideration.

Norman A. Scotch • Boston University School of Medicine, Boston, Massachusetts.

3. A component of the American class system is social mobility.
   To be sure, such mobility is not as great as some commentators
   believe. But it is sufficient to add complications to self-
   identification in terms of class membership, as well as ambivalence about the values predominant in one's particular class.
   For many people, this ambivalence produces inconsistent attitudes and practices regarding parent-child relationships, as
   well as those between wives and husbands.
4. The most important aspect of Dr. Kohn's paper was the implication of his observations for providers of services to families.
   These are of utmost importance.

Dr. Kohn points out that values, roles, child-rearing practices,
world views, indeed perceptions of reality are powerfully influ-
enced and shaped by social class membership. Social class membership, he reminds us, is the best single predictor of human behavior in virtually every sphere: from political ideology to sexual
preferences and from economic consumption to life expectancy.

Moreover, the concept of social class not only permits us to
describe and predict the behavior of members of particular classes
with surprising accuracy, but it also permits us to predict how society in general and social classes in particular will view the behavior of members of *other* classes. That is to say, we can predict
that many upper-middle-class members will view with dismay,
amusement, condescension, or perhaps envy the behaviors of
lower- or working-class members. Thus, "lower-class people do
not plan ahead—or have too many children—or are too dirty—or
don't invest enough in children—or are not too intelligent," and so
on. Likewise, working-class people may view the values and behaviors of middle classes as "too permissive—or too effete—or too
restrained—too fancy", and the like. Please note that the accuracy
or inaccuracy of such views is less important than the fact that *there
are fundamental differences* in perception. (Also please note that here
we are talking about simple ideal family forms and not yet about
variations that will be discussed later.)

Do such differences in behavior and values influence or affect
the practice of medicine, psychiatry, social work, or other helping
professions?

Yes, of course. Until *very* recently the preponderance of health professionals tended to be recruited overwhelmingly from the middle and the upper-middle classes. And this tended to make for rather good rapport when such health professionals were treating patients from the same social classes. There is little doubt, however, that when interacting with members of other classes—more specifically the lower classes—problems arose.

Most clinicians do not have access to the very important points to which Professor Davis directed our attention earlier—that the family is a social institution created not in Heaven but by man; an institution shaped by numerous historical forces and, in turn, shaping other institutions. The result is a complexity of human functions, values, and activities often in flux, often changing. Nor do such clinicians often have the benefit of Dr. Kohn's analysis of some of the fundamental differences between classes that are so important in shaping behavior.

No—what clinicians most often have instead is their own membership in a particular class and the world view such membership gives them *and–more importantly*–the sense that this world view is the proper or correct view. This presents problems for professionals. Often they simply do not understand the source of such differences and, unfortunately, see them in evaluative or judgmental terms.

For the patient or client of such professionals, there are problems of considerable magnitude ranging from simple lack of communication to the provision of inferior services. But it is a mistake to think that the resultant problems of class distinctions fall only on the patient. For a measure of how good a person is in his or her profession is how *effective* he or she may be. Those who are most dominated by their own values have a measurable tendency to be less effective when dealing with people from classes different from their own. This is *their* problem and not the problem of the client or patient.

These few words are, of course, an oversimplification of the actual state of affairs. There are people who can transcend their own class membership and who can deal effectively with others. Moreover, significant numbers of the health professions have been upwardly mobile and can understand the world views of different social classes—and these numbers are increasing. Likewise, there are core values (honesty, decency, and other universal values) that all classes share and that continue to permit effective interaction. Furthermore, the training of professionals increasingly includes attention to the validity of different life-styles and to the importance of techniques that increase effectiveness across class barriers, and this new training gives us considerable optimism.

2

Practical Considerations

Ira D. Glick

Dr. Kohn's chapter on social class and parental values and practices
is a comprehensive and clearly written review of a large body of
literature. The main thesis is that there are substantial differences
in how parents of dissimilar social-class positions raise their children and that these differences in parental practices result *in the
main* from the class differences of parents' values for their children.
These parental values, in turn, result from varying conditions of
life experienced by the parents.

The issue now becomes how to use these very interesting data
and translate them into clinical diagnostic and treatment language.
Let me focus more specifically on family psychotherapy rather than
on clinical practice more generally. First, some background. The
family therapist starts from a model that views the family as a system. The identified patient's symptom is viewed as a symptom of a
disturbed family system.

As far as *diagnosis*, the paper provides information on how

Ira D. Glick • Cornell University and New York Hospital Medical Center; Payne
Whitney Clinic, New York, New York.

"normal" families function. The next step for the clinician is to compare this to the family in treatment. That is, in order to decide whether or not to change your family, you first figure out how the family compares to others of a similar socioeconomic class and culture. The second point the paper illustrates is that values and practices can be generalized by class. Of course, the converse is not true, that all families in a given class have the same values and practices.

Once you have made a diagnosis, the next step is family *treatment*. I come from a bias of using family-system concepts in conjunction with psychodynamic theory to understand families. Looking at Dr. Kohn's material, let us make six clinical points about changing families. First the clinician must compare the data on values and practices from his family with those of other families from the same social *class*, that is, compare child-rearing practices, ideas about education, etc. Second, the clinician must integrate the data in the context of that family's *culture*; Chinese families have different values and practices than do black families, who in turn differ from white families. Third, the clinician must separate an individual family value or practice from those emanating from *changes in society*. I believe Dr. Davis's chapter touched on this; societal changes in values and practices have differing effects in the family. A good example is that although most custody decisions are now made in favor of mothers, there is a long historical tradition of custody going to fathers for quite different reasons. The fourth area to evaluate is *individual differences*. What are individuals in this particular family like and what have their backgrounds been? Individual and background variance will substantially influence child-rearing and educational practices. I'm reiterating a disclaimer that Dr. Kohn has made, but I want to make it clear that a variety of influences operate. Fifth, the clinician should *not stereotype families*. It's one mistake to think that there are no common class values, and the opposite mistake is to stereotype all families; stereotypes and myths exist for families of all classes and all cultural backgrounds. Finally, the sixth area the clinician should try to evalute is the *real-life constraints* on families, such as economic conditions. Such constraints are a major problem in single-parent households.

What the clinician tries to do is understand the family, then try interventions to change the family, and finally, help them cope more effectively with their situation. I think the central issue is to try to work with (rather than against) the family around their particular class values, cultural values, and cognitive system. If parents have certain beliefs about child-rearing practices, it is *very difficult* to get them to make major changes in those practices. The usual mistake, and failure of family therapy, in my experience, has come about when a therapist tries to impose his/her own class values—and I underline the word *values*, not facts—on a particular family. You have to start with the *family* values and modify them in as helpful a way as you possibly can. For example, a common problem for therapists is the family that is "constrictive" and modifies behavior in an "authoritarian fashion." The family therapist will try to get the parents to switch their methods, for example, from spanking to talking.

Another way to change families is to reintroduce children *back into* their culture. One couple I recently saw were reintroduced to their culture by such activities as returning to church. In families that I treat, we encourage the children to work with some *neighborhood organizations,* and that has been useful because rather than fighting the parents, they become involved in a more socially useful kind of sublimation. Finally, it must be emphasized that the family therapist must *respect* individual dissimilarities in values and practices.

Let me illustrate that with one case. In a Chinese family that I see, the identified patient was a 19-year-old boy with a working diagnosis of schizophrenia. He had not gone to school for three years, had not talked in a year and a half, was quite delusional, and was starving himself. His mother and father had been born in China and had been in this country about 20 years. The father was an auto mechanic, the mother a bookkeeper. There were two older sisters: the eldest was shy, quiet, and "good"; the younger was "rebellious."

I'd like to add some interesting facts to give you the flavor of the family. First, when the identified patient (this 19-year-old boy) was born, his mother quit her job as a bookkeeper, deciding what she needed was to fulfill herself by staying home to raise her son. Second, the parents stopped their sexual relationship; they had not had intercourse in the last 19 years. In addition, the two sisters were both dating the same man, an auto mechanic like the father; although they both had vague inklings that this might be happening, it had not been discussed.

The first priority was to decrease the identified patient's symptoms with medication so that he wouldn't kill himself or starve himself to death. He was given phenothiazines, which resulted in a moderate decrease in his symptoms. It didn't solve a lot of other problems, but it at least kept him alive, so we could start family therapy. The next goal was to change the parental coalition; we started marital therapy in conjunction with family therapy. The third order of business, and this came later on, was to start sexual therapy. A behavioral modification program was used starting with mutual back rubbing, which resulted, nine months later, in twice-weekly intercourse. This was after a 19-year absence of sexual activity.

Once that was accomplished and the parental coalition was strengthened, we encouraged this couple, who had not done anything of a social nature, to go to a Chinese movie with the identified patient. They had done nothing by themselves for 10 years and I didn't want to cut that off suddenly. After the parents built a good relationship, we then started two rehabilitative measures with the identified patient once his symptoms decreased. The first was a sheltered workshop to help him achieve a sense of gratification and build up his self-esteem. Second, he was started in a socialization group with people from his workshop. Simultaneously we sought to improve the functioning of both sisters by meeting separately with them. This took a long time. It was accomplished when the shy daughter moved out of the house and began dating other boys. This was connected to the father's wanting to "hold onto" the shy daughter because he wasn't getting gratification from his wife.

That's a quick synopsis of a year and a half's work, to highlight the kind of diagnostic and therapeutic issues involved in working with families with different class and social values. These people had very fixed ideas about how things should be done in terms of child rearing. All their ideas were based on the traditional Chinese values; that is, "the older son should be the boss, the favored one, who gets everything." It absolutely "threw" them when the older son was really the least capable in the family and became the one with the most severe problems.

3

The Adaptations of Urban Black Families:
Trends, Problems, and Issues

Albert J. McQueen

The theme of this conference—is contemporary family dying or developing?—yields seemingly paradoxical results when applied to black families. For many years, the conventional wisdom of social science held that black families were either dying or had attained a precarious equilibrium of perpetual existence in the shadows of death and decay. Over 300 years of survival suggest that the announcement of the death or near-death of black families has, in the words of Mark Twain, been premature and grossly exaggerated. And yet, on the face of it, 300 years does seem like an awfully long time to be still talking about the development of black families. There is, of course, no paradox here. Instead, we are confronted with a complex of social conditions and behavioral responses that social scientists have only recently begun to under-

Albert J. McQueen • Oberlin College, Oberlin, Ohio.

stand and interpret realistically as experienced by black people themselves.

In this paper, I shall explore some of the complex dynamics of black family life with three broad aims in view. The first is briefly to summarize the dominant social science images and interpretations of black families. The second aim is to discuss certain major characteristics and trends evident among black families in recent years and some of the underlying reasons for them. The last task is to analyze family functioning among inner-city poor and near-poor people, with particular reference to the strategies by which they cope with the material deprivations and dangers inherent in their environment, seek to maintain viable and satisfactory family life, and endeavor to raise their children for a better future.

SOCIAL SCIENCE IMAGES

Turning to the first area, two quite coherent, diametrically opposed viewpoints can be delineated. The first, which dominated social science until the mid-1960s, is here labeled *pathological adaptation*. The second orientation was a reaction against the first and is currently very much in evidence. Let us call it *healthy adaptation*.

Before discussing these two approaches, it is important to understand why the word *adaptation* occurs in both. It does not seem appropriate here to speak of *adjustment,* which implies short-term changes in persons and groups to achieve a good "fit" with the social environment. Historically the socio-cultural environment of black Americans, far from being supportive and encouraging of positive family adjustment, has been permeated with oppressive forms of racism and thus greatly inimical to viable family life. Family survival and the achievement of family goals have over the years required sustained struggle against powerful societal forces and efforts to devise social stratagems that could yield positive results. *Adaptation* here refers to developments and changes in family structures and functioning in response to persistent problematic social conditions, as persons seek to realize and consolidate values that give substance and meaning to family life. Social scientists have disagreed widely about the nature of black family adaptive patterns, as well as in their interpretations of the sociocultural meanings and consequences of such patterns.

Pathological Adaptation

For the sake of brevity, I shall take note of only two publications, a quarter of a century apart, that were key statements of the image of pathological adaptation. The first, E. Franklin Frazier's *The Negro Family in the United States*, published in 1939, entrenched this viewpoint, and the second, Daniel Patrick Moynihan's *The Negro Family*, appearing in 1965, sparked the reaction that resulted in its demise. Frazier was very much aware of the emergence of strong and stable families in Afro-American experience. But he saw the dominant pattern as characterized by pathology. Simply stated, his position was that slavery destroyed African cultures and family patterns and stamped a maternal character upon black families, leaving the mother-child bond the most durable and meaningful feature of family life. Conditions under slavery undermined efforts of males to support and protect their families and to establish authority in them. Worst of all, in Frazier's view, is that blacks did not have ample opportunity to become acculturated to the dominant nuclear family norms of the society. The massive migration of blacks out of a simple rural folk culture into highly competitive and racially discriminating Northern cities that began prior to World War I could only mean additional trouble for family life. The predictable result, in short, was a further weakened family structure characterized by conditions of social disorganization such as broken homes, sexual promiscuity, illegitimacy, crime and delinquency, welfare dependency, and desertion.

Moynihan built his study directly on Frazier's interpretation. He drew together mounds of statistical data from U.S. Census Bureau and other official reports in an attempt to demonstrate that black families, except for middle-class ones, are "matriarchal," caught in a "tangle of pathology," disintegrating," "crumbling," etc. While Moynihan stood out for couching his interpretations in highly dramatic and pejorative terms, many other social researchers and commentators have contributed significantly to the image of dysfunctionality. From numerous writings, I have culled four basic themes denoting pathology.[1]

The first is that black families, because of broken homes and the dominant role of females (matriarchy), are unstable, structurally weak, socially disorganized, and handicapped in carrying out essential family functions.

The second theme is closely related to the first. It suggests that black families have little capacity to improve their lot in society because they are seriously deficient in needed social resources and competencies. James S. Coleman, for example, assumes an extreme view in asserting that family defects "prevent the growth of economic and socialization resources" for social mobility (1971, p. 43):

> It is in family resources that blacks have probably the greatest liability. They are both cultural—the absence of cultural traditions that guide and assist parents in socializing their children—and structural. The structural liabilities consist in the absence of a strong male role in the family and the number of families in which the father is absent entirely or only intermittently present.

The third theme, best formulated by Lee Rainwater (1966), involves the notion of self-victimization. Existing under persistent conditions of extreme duress causes family members to resort to adaptive strategies of social psychological survival that produce expressive and violent behavior within the family, as well as tendencies toward depression and fatalism.

The last idea is that the situation of black families is growing steadily worse, particularly as revealed in the increasing proportions of female-headed households and widespread illegitimacy.

This decidedly negative imagery, especially as expressed by Moynihan, provoked sharp reactions among a considerable body

---

[1]Besides Frazier and Moynihan, several of the authors whose ideas are reflected in these themes are Bernard (1966), Blood and Wolfe (1960), Etzkowitz and Schaflander (1969), Coleman (1971), Pettigrew (1964), and Rainwater (1966, 1970). It should be noted that many social scientists who dwelled upon pathology did so with a sense of outrage and moral indignation about the destructive consequences of racial exploitation upon the black family. Others undoubtedly harbored essentially pernicious motives.

of social scientists, black and white.[2] As they critically appraised major works reflecting this orientation, numerous flaws were unearthed in the selective use and misinterpretation of statistical data; the espousal of exaggerated value-laden positions without adequate empirical data; overgeneralization; and neglect of the positive and functional aspects of subcultural values and social patterns.

Andrew Billingsley's *Black Families in White America* (1968) was a notable indictment of the pathology vintage of research and theory, as well as an attempt to formulate a conceptual framework for more realistic appraisals of black families. He discredited the prevailing method of interpreting black family phenomena by almost exclusive reliance upon comparisons of black and white family data, with the latter always providing the norm of desirability. If we are to move beyond superficial analyses, according to Billingsley, then we must study black families in their own right as adaptive systems with varied structures, modes of functioning, and consequences for family members. Also, if we are to understand the values and meanings behind observable family phenomena, we must seek to comprehend the nature of Afro-American culture as a set of behavioral, symbolic, and material patterns that is the foundation for life in black communities.

Historians have played a key role in reinterpreting the nature of black families. The long-accepted thesis that slavery virtually wiped out black families and that blacks did not develop strong family norms has been decisively disproven by the seminal works of historians such as Eugene Genovese (1974), George Rawick (1972), John Blassingame (1972), Herbert Gutman (1976), and others. Family sentiments during and after slavery were deeply ingrained, and men and women struggled mightily against great odds and with considerable success to keep family units together.

---

[2]As controversy grew around interpretations of black families, a vigorous phase of critical analyses and research ensued. Some of the many commentaries and studies I've found valuable are Hyman and Reed (1969), Ladner (1971), Lewis (1967), McWorter (1969), Farley and Hermalin (1971), Liebow (1966), Turner (1972), Staples (1971b), Stack (1974), and Heiss (1975).

Healthy Adaptation

The counterimage of healthy adaptation emphasizes the kinds of qualities that made endurance and survival under hostile conditions possible: "resilient," "ingenious, "long suffering," "realistic," "resourceful," and the like.[3] Naturally family casualties and failures have occurred and will continue to occur, but from this perspective they should not be allowed to detract from accomplishments of heroic proportions under the most trying circumstances. One of the most systematic discussions of the qualities that have made this possible has been made by Robert B. Hill in his little book, *Strengths of Black Families*. While not claiming his list of five basic strengths to be comprehensive or exclusive to blacks, Hill endeavors to demonstrate their prevalence empirically. They are:

1. Strong kinship bonds. Extended family sentiments and ties tend to be maintained, particularly among poor people. They are manifested in a willingness to aid relatives in need and to accept both kin and nonkin children and adult relatives into the family.
2. Strong work orientation. This is revealed in attitudinal studies and the high proportion of working wives in husband-wife families.
3. Adaptability of family roles. This primarily involves equalitarian decision-making in family affairs, which has been found to be more characteristic of black families than the popular image of matriarchy or wife-dominance.
4. High achievement orientation. The desire to achieve in education, occupation, and income tends to be pronounced among parents and their children, though it often is not matched by an equally strong sense of efficacy or optimistic expectations of success.
5. Positive orientation toward religion. While not all blacks are religious any more than any other segment of the population,

---

[3]Billingsley's book (1968) had a major impact in helping to shift social science thinking from preoccupation with dysfunctionality to a recognition of the positive, functional aspects of black family adaptations. The image of black families as strong and able to endure in harsh environments was concretized in his rhetoric and arguments.

the church is a central institution in the black community, a bulwark for the maintenance of family values of respectability, perseverence, and achievement.

This cursory view of pathological and healthy images of family adaptation might well dispose the reader quickly to choose sides. Even though I definitely tilt toward the position of healthy adaptation, I think it would be a mistake to deny or ignore weaknesses and disabilities that may be associated with certain family types and behavioral patterns. Many families cope surprisingly effectively under conditions of racism, poverty, or powerlessness. At the same time, however, all is not well in myriad others exposed to these scourges. The main interest here is to understand why and under what conditions families function at different levels of effectiveness in meeting family needs and realizing their values and goals. This requires study of family coping strategies, the topic that is central to my research on family functioning. This area will be explored after a discussion of certain characteristics and trends among contemporary black families.

FAMILY CHARACTERISTICS AND TRENDS[4]

In 1974, there were, according to the U.S. Census, 5.5 million black families and 24 million black people in the United States. In this overwhelmingly urban population—over 80% resided in urban areas—roughly three out of every five persons were concentrated in inner-city sections of our largest cities. This means that the majority of black families lived in environs that have been characterized by the economist Daniel Fusfeld (1968, p. 75) as existing in a "condition of permanent depression":

> Even though the rest of the economy may be prosperous, even booming, and may feel the inflationary pressures of rising aggregate demand, unemployment in the urban ghettos will remain high. Ghetto unemployment rates stay at levels which would signal a serious depression if they prevailed in the economy as a whole.

Associated with this permanent depression, unemployment rates among blacks consistently have approximately doubled the rates

---

[4]The factual materials in this section are so numerous that most will not be referenced. The sources for the facts and figures herein are the U.S. Bureau of the Census and the National Urban League documents listed in the bibliography.

for whites during the past three decades. Daniel Moynihan (1965a, p. 748) has spoken of this fact as "one of the seemingly fixed ratios of our economy." In the third quarter of 1976, when unemployment (official government rates) among whites was 6.9%, it was 13.2% among blacks. The National Urban League's index of hidden unemployment, involving "discouraged" workers no longer seeking employment and those who take part-time work involuntarily, estimated a 22.4% unemployment rate among blacks during the same period (1976, pp. 2-4). Official unemployment among black teenagers was 36.1% during that period and 15.6% for white teenagers.

Although significant occupational and income gains were made by blacks during the 1960s, the fact remains that the black community is still fundamentally poor. In 1973, before the 1974-1975 depression, 28% of all black families, by the Bureau of Labor Statistics standard of $4,540 for a nonfarm family of four, were designated as poor. The National Urban League, recognizing inadequacies in this standard and setting a higher poverty threshold of $6,810, arrived at the figure of 44% of all black families as poor (*Special Policy Report*, 1975, p. 24). Even if Urban League standards are considered too liberal a cutoff point for poverty, we would still be justified in considering the 44% figure a reasonable measure of the *poor and near-poor* among black families.

Rather than attempt to unravel the complex interactions of economic conditions and family structures and functioning, the remainder of this section will be concerned with selected family characteristics and a discussion of factors that might explain them.

The first point to recognize is that black families are not fundamentally different from American families in general. But distributions of characteristics and types diverge from white family statistical norms in several respects, and important differences exist in cultural and behavioral patterns. For example, there are lower rates of marriage and remarriage among blacks than among whites. Also, on the average, blacks tend to be somewhat older when they marry. Following a national trend, the median size of black families has dropped since 1960, but it was still somewhat higher in 1973 (3.9 persons) than the size of white families (3.6 persons). The majority of black families are nuclear families headed by males, and the majority of black children under the age of 18 live with both parents. Although extended family sentiments and interactions remain strong, few black families—6.4% in 1970—tend to include three or four generations.

At a time when American families are undergoing rapid changes and signs of instability are very much in evidence, the proportion of black families that do not conform to dominant nuclear family patterns and that have experienced dissolution has grown significantly. In the first place, there has been a sharp drop in husband-wife families among blacks, from 74% in 1960 to 61% in 1975. During that same period, the proportion of black families headed by females jumped from 22% to 35%. White families, 87% of whom were husband-wife families in 1975, had changed in the same directions, but by only one or two percentage points in each case. Between 1960 and 1975, black female family heads decreased in median age from 43.8 to 37.7 years, while the percentage single doubled (from 11% to 22%) and the proportion separated and divorced increased from 42% to 50%. Just over 71% of the black women heading families in 1975 had children of their own under the age of 18 living with them. In 1974, 53% of black families headed by women were officially designated as poor, compared to 14% of those headed by males.[5]

For reasons too numerous to discuss here, illegitimacy statistics tend to be of dubious reliability.[6] The available data, however, reveal that illegitimacy rates are six times higher among blacks than among whites. The white rates, however, have just about tripled during the years 1960-1973, while the black rates nearly doubled during the same period. In 1972, 40% of all black births apparently were illegitimate. As we should expect from the foregoing data, the

---

[5]There has been a significant increase in the *number* (in contrast to the percentage) of poor black families in the 1970s. This can be accounted for solely by increases in female-headed families, since the number of male-headed families in poverty has decreased. The National Urban League (1975) estimates that 70% of female-headed black families were poor in 1973 and that the numbers of such families have increased more sharply than official figures indicate. See *Black Families in the 1974-75 Depression*, p. 10.

[6]For discussions of problems of understanding and measuring illegitimacy, see articles by Lewis, Herzog, and Ryan in Rainwater and Yancey (1967) and also Chapter 4 in Ryan (1971).

proportion of black children under 18 years of age living with both parents has declined: in 1960, the figure was 67% and in 1974, 52%. There are similar trends for white children, but again at much lower levels. A very large proportion of black children live either with their mothers or other relatives or have been adopted by nonrelated persons.

Taken at face value, these patterns and trends would seem to confirm the pathological adaptation view discussed above. Before embracing such a conclusion, however, it is important first to understand why black families are changing in these ways and then to investigate actual adaptive behaviors.

The following interpretation is proposed for the disturbing shape of events that seem to have overtaken black families. There has been an increase in unmarried persons, particularly women, and troubled families for three main reasons: (1) economic distress, (2) a shortage of black males, and (3) changing norms for marriage and family life.

Economic Distress

Black educational, occupational, and income gains that were achieved in the 1960s have not been reversed in the 1970s, but there have been significant losses. Since 1960, the proportion of employed black males 16 years of age and over has steadily declined from a high of roughly 74% of those in the labor force to a low of about 59% in 1976. Black family income has fluctuated between 60% and 65% of white family income during the past decade. But more important, the median family income for blacks, in constant 1975 dollars, generally stagnated and, indeed, had fallen slightly by 1975. Also the proportion of black families with two or more earners has declined somewhat in the 1970s. These conditions, when coupled with the high levels of unemployment and poverty discussed above, yield a picture of increasing economic distress for large numbers of black families.

The effects of these kinds of economic changes on families have been well documented in the monumental University of Michigan longitudinal study of a single national sample of families covering the five-year period 1969-1973. Robert Hampton (1975) found that family income was a powerful predictor of marital disruption for both black and white families. Marital stability decreases as income goes down. In their impressive book, *Time of Transition: The Growth of Families Headed by Women* (1975), Heather Ross and Isabel Sawhill, drawing upon the same research, succinctly summarize the main economic factors they found to be associated with family instability (p. 19):

> The nonwhite couples were found to have almost all of the characteristics associated with marital instability in much high proportions than the white couples. They were more likely to have low-earning husbands and few assets, to live in the central cities of large urban areas, to have working wives, and to have income more evenly split between husband and wife. Most importantly, they reported much more unemployment in recent years and this was the single most important variable accounting for the higher instability among nonwhite families. After controlling for these and other differences between the two groups, there were no statistically significant differences in separation rates by race.

Vast numbers of poor and near-poor black families, as well as the many middle-income families whose standard of living is precariously maintained by both husbands, and wives, working, are caught in a triple bind of inflation, unemployment (or the threat of it), and racism in its many overt and subtle manifestations.

Shortage of Black Males

This reason for a high rate of female-headed families among blacks has been largely neglected by social scientists. For the entire population, the sex ratio in 1975 was 95 males for every 100 females. For the black population it was 91 males for every 100 females. Jacquelyne Jackson (1971) has demonstrated that the black sex ratio has declined throughout most of the 20th century, from a high of 99 in 1900 and 1920 to the 91 figure for the 1970s. When we consider the most marriageable ages for men and women, 15-44 years, the sex ratio is even less favorable for establishing families: 89 males for every 100 females (in 1974). Black women face what Paul Glick, veteran population expert with the Bureau of the Census, has described as a severe "marriage squeeze" (1975, p. 17). A simple example will show how it works. Let us assume that black males and females of different ages were distributed randomly among cities in this country. If there were 100,000 women aged 15-44 in a particular city, there would be 10,000 fewer men of comparable ages. Even if economic conditions did not mitigate against stable family life, many women obviously would not be able to have husbands, unless the marriage norms and social patterns were changed to facilitate either interracial marriage or polygyny.

The most important reason for this shortage of black males is evident in life expectancy (at birth) trends during the past 25 years (U.S. Bureau of the Census, 1976, p. 60):

	1950	1960	1970	1974
White females	72.2	74.1	75.6	76.6
White males	66.5	67.4	68.0	68.9
Black females	62.9	66.3	69.4	71.2
Black males	59.1	61.1	61.3	62.9

While the low life expectancy of black males increased at a faster rate (6.4%) than that of white males (3.6%) and females (6.1%) during that period, it was far surpassed by the growth rate of black female life expectancy (13.2%) and was over eight years lower than theirs in 1974. This widening survival gap suggests, in effect, that the shortage of black males can also be seen as a growing surplus of black females.

No effort will be made in this paper to explain these life expectancy discrepancies. It should be recognized, however, that they do not entirely account for the current low sex ratio among blacks. This phenomenon is in part an artifact that stems from a greater tendency for males to be undercounted in censuses than females. The Census Bureau has estimated that black males were undercounted in the 1970 census at the rate of 8.9% and black females 4.9%.[7] If corrections are made for these estimates in computing the

---

[7]These figures are taken from the National Urban League document *Estimating the 1970 Census Undercount for State and Local Areas* (1973). The Research Department of the League has developed a tentative formula based on official undercount estimates by age, sex, and race for correcting population counts for cities and states.

1970 sex ratio, it would increase to 94 black males for every 100 black females instead of 91. Even if this new sex ratio figure were a reasonably accurate estimate, its significance for rates of marriage would still not be readily apparent. The effects on families, largely negative, of conditions where there is relatively high black male involvement (for example, unemployment, enlistment in military service, participation in marginal and illegal commercial ventures, and imprisonment) would have to be investigated for a fuller understanding of family characteristics and trends.

Changing Norms for Marriage and Family Life

In their perceptive study of the growth of families headed by women, Ross and Sawhill (1975) hypothesize that elevating the social and economic status of women has loosened many constraints of traditional family life and enabled women to move in and out of marriage more readily in search of personal satisfactions. Thus, they see the female-headed family as a transitional phenomenon of "relatively short duration between one traditional family structure—usually a husband-wife family—and another" (p. 6). Their thesis, which has substantial data support, suggests that family instability may not necessarily mean the erosion of the nuclear family unit. Instead, it may mark the emergence of a set of cultural norms that allows for greater experimentation and adaptability in family living arrangements. What seems threatening or unfortunate now may at a later date seem normal and useful.

Ross and Sawhill are cautious in applying these ideas to the growth of female-headed families among blacks. One of the prominent confounding factors is the continuing precariousness of male employment, which we have seen is a key variable in marital instability. Another factor may be the greater dependence of black women on welfare, which seems to create "an incentive for them not to marry or report their marital status accurately" (p. 88). Their interpretation suggests that black female-headed families may represent less of a transitional pattern based on resources and options and more of a condition imposed by economic and social constraints. It is not farfetched to hypothesize, however, that with increasing socioeconomic status and urban residence certain segments of the black population have the freedom and are subject to influences that engender more flexible orientations to marriage and family arrangements. Just as the number of black "unmarried couples" has been reported to have grown sixfold between 1960 and 1970 (Glick, 1974), some unknown portion of the growth of black female-headed households may be accounted for by a greater tendency to move in and out of marriages by choice in search of more satisfying interpersonal and living arrangements.

Concluding Comments

The meaning and consequences of black family instability are not at all self-evident. Single-parent families and bearing children out of wedlock are not in their nature pathological. Many observers have pointed out that extended family supports act as quite effective cushions against disaster for poor and near-poor families. Carol Stack's fine study of female-headed families on AFDC welfare, *All Our Kin* (1974), graphically portrays the workings of various forms of mutual aid among relatives as vital factors in day-to-day survival and the achievement of some measure of well-being. Exchange arrangements are especially important in the care and raising of children, many of whom are illegitimate. While having children out of wedlock is not positively valued, having children is.[8] The community tends not to place a stigma upon mothers or their children in such cases. The children become regular family members and often are reared by realtives, especially grandmothers, kept for periods of time by relatives and friends, adopted outright, or given some support by the fathers or their relatives.

Since black family instability stems largely from economic deprivations, demographic imbalances, and perhaps changing social norms, we should not expect it to be a socially inherited phenomenon as assumed by proponents of the once-popular notion of a "cycle of poverty." Indeed, what little evidence there is supports the idea that family instability is not to any significant degree passed on to younger generations. Large-scale studies by Heiss

---

[8]See Ladner's insightful study (1971) of role models for the socialization of black adolescent girls in low-income families. Also Stack (1974).

(1972) and Pope and Mueller (1976), while far from conclusive, reveal that there are only small and inconsistent relationships between instability in the parental generation and the dissolution of marriages among male and female respondents. Just as external conditions must bear the weight of responsibility for creating family instabilities, the evidence suggests that significant improvements in these conditions should be followed by more stable family patterns.

COPING WITH POVERTY

If poor inhabitants of inner-city ghettos had the power and know-how to restructure the social institutions that control opportunities and resources and that relegate them to the bottom rung of the status system, the lessons of history are that they would do so. But the harsh reality is that theirs must be a daily struggle against material scarcity and deprivation and for some measure of self-respect, dignity, and improvement in the quality of their lives. The degree of their success or failure is a function of external forces that are essentially beyond their control. But not completely. The research I shall report on was designed to explore the kinds of orientations and strategies by which poor black families cope with exigencies and try to exercise some control over their destinies. It assumes that the broad configuration of Afro-American culture comprises a variety of values, symbols of status and worth, instrumental means to achieve goals, life-styles, oral and written traditions, etc., which are historically derived alternative "solutions" for resisting and rising above oppression and degradation.

This research was carried out in the mid-1960s in the Cardoza area of the District of Columbia.[9] The sample consisted of 100 randomly selected poor and near-poor families with 14- to 17-year-old children in them. Only the mothers were interviewed.

The study sought to identify two types of poor families: those who ranked high on measures of family well-being, that is, those who were able to take care of basic family needs and realize impor-

[9]This study was carried out during the summer of 1965, when I was a member of the Center for Youth and Community Studies, Howard University.

tant family values, and those who ranked low. The former group would be the effective copers and the latter the ineffective copers. The research objective was to determine the social characteristics and coping strategies that distinguished between the two groups. In order to measure family well-being, the mothers were asked questions in five areas: how adequately the children were fed, how adequately the children were clothed, whether the rent was paid regularly and on time, whether the family owned or was buying a home, and whether the family had savings. Sharp differences emerged among the families, which made it easy to delineate two types.

Measuring the poverty level was too complicated to explain in this brief paper.[10] The important point is that a poverty index was determined for each family, based on the actual family income divided by the amount of income necessary to meet basic family needs. A 1.0 index means that the family has just the amount of income they need—they are near-poor—while a 0.5 index would leave no doubt that the family is poor, since they would have only half the amount of income required to take care of basic family needs.

This analysis will dwell upon the 57 families who were classified as poor by virtue of having poverty indexes below 1.0. The 34 effectively coping families and the 23 ineffectively coping families averaged between 60% and 65% of the yearly income they needed. The ineffective copers I have labeled *troubled poor* families because so few fed and clothed their children well or paid the rent regularly and on time. The effective copers, on the other hand, ranked very high on these variables and had overall profiles that made the label *future-oriented poor* appropriate for them.

First, a word about some differences between these two types. All the families tended to be large, with about seven persons in the household, but the troubled families were somewhat larger, and more of them tended to include relatives and nonrelated persons. The future-oriented families were more likely to have male heads: 85% compared to 65% for troubled families. Although a somewhat

---

[10]The source for formulating and computing the poverty index is the welfare index devised by Morgan, David, Cohen, and Brazer (1962, pp. 188-191).

larger percentage of the male heads in the future-oriented families were unemployed, three times more wives in these families worked than in the troubled families. Just over a fifth of the troubled poor families received some form of welfare, compared to 6% of the other type.

Themes in Coping Strategies

What kinds of orientations and strategems make it possible for so many of these very poor families to thrive as stable units and cause some to have much higher capabilities to manage their affairs than others? My interpretation of the data has yielded five basic themes, perhaps better called hypotheses, that suggest answers to this question. The families will be compared on each in turn.

Strong Family Orientation. When asked what kinds of things they and their spouses (where present) would like to accomplish for themselves during the next five years, the mothers in both types of families responded strongly in terms of family-related goals: improving or having a better home, living in a better neighborhood, fulfilling material needs for the family (furniture, food, etc.), raising children well, having better family relations. The personal goals related to self-improvement: getting more education and having better health, for example. Judging from their effectiveness in attending to family needs, the future-oriented families apparently actually devoted more of their resources to family goals, thus following through better on the kinds of goals they shared with troubled families.

Mobility Aspirations. At no point in the questioning did any of the mothers suggest socioeconomic-status mobility aspirations for themselves or their husbands. To improve the quality of life in the family seems to have been their main preoccupation. When asked about educational and occupational aspirations for their sons, however, mobility goals were clearly articulated, either in terms of being able to have a reliable occupational and income position at the working-class level or moving up into professional, business, and other white-collar occupations. Mothers in future-oriented families expressed the higher aspirations; for example, 40% wanted their sons to complete a college education, compared to

26% of the mothers in troubled families. Likewise with occupational aspirations: 63% of the former aspired to upper white-collar occupations for their sons to 40% of the latter. Perhaps of greater importance is the fact that far more mothers in future-oriented families actually expected their sons to accomplish the goals they held for them: a third of them expected their sons to finish college in contrast to 11% of the mothers in troubled families. Close to 70% of the latter expected their sons only to finish high school, if that, while only 30% of the future-oriented mothers held the same view. For whatever reasons, future-oriented mothers conveyed a strong sense of efficacy and confidence regarding the probable achievements of their sons. It is reasonable to expect that they would be more attentive, encouraging, and supportive of their sons for positive achievements than mothers in troubled families, who seemed far less hopeful and confident.

Quest for Respectability. In his book *Blackways of Kent* (1955), Hylan Lewis described families along respectable and nonrespectable dimensions, in part in terms of their orientations to and relations with the two pivotal institutions in the black community, the church and the tavern. While such a classification is surely a great oversimplification of life in the black community, it can be useful for understanding coping behavior in families. There is no direct evidence that troubled family members were more active in the "good-time" life of the dance hall or the tavern or that they were less concerned about projecting an image of respectability in the community. What the evidence does make clear is that the wives and the husbands in the future-oriented families geared their lives to the church strikingly more than their counterparts in the troubled families, whether in terms of church membership, attending services, or membership in church organizations. In each category, the wives were more involved than their husbands. Greater concern of future-oriented families for respectability through religious activities thus seems to be a logical hypothesis. It also suggests that parents have their family values and aspirations reinforced through the church and reap the benefits of a supportive community for the kind of asceticism required for hard work and disciplined management of family resources and affairs.

Planning. The preceding idea about the behavioral consequences of church participation becomes more cogent when we realize that 35% of the future-oriented families owned or were buying their homes and one-fourth of them had some savings, despite their markedly deficient incomes. Unless we assume that they had been beneficiaries of unusual good fortune in the past, the idea is compelling that these parents must have exercised unusual diligence and discipline in planning and acting to improve their family situation. Only one troubled poor family was buying a home, and none had savings.

Self-Reliance. The numbers here are small and demand caution in drawing conclusions, but they do suggest a remarkable fit with the pattern that has unfolded. One-third of the mothers in the future-oriented families and two-thirds of those in the troubled families stated in response to a question that there were times when no money was available to meet basic family needs. When asked what actions are usually taken, all of the future-oriented mothers said they try to "make do" with what they had, make out somehow until the next payday; only 10% said that they borrowed from relatives or friends.[11] Mothers in troubled families stated just the opposite: 80% tended to borrow from relatives or friends, and only 43% tried to "make do" on their own. The interpretation I propose is that troubled poor families are much more likely to be involved in extended-family mutual-aid networks of the sort that Carol Stack describes in *All Our Kin* (1974). This system of reciprocity can be an extremely effective means of pooling resources to provide help to those ordinarily in need, as well as assisting with the recurring crises that plague the poor. Of course, there are always persons in one's network in need, and a person is obligated to help if he can. This means that it is virtually impossible to have control over one's resources, to manage the family income for optimal advantages in the quest for family goals. It is for this reason, I hypothesize, that future-oriented parents tend to avoid extended-family reciprocities and obligations. They would not be able to plan and allocate their financial resources for the fulfillment of their goals to any appreciable extent. This does not mean that future-

---

[11]Each interviewee could mention more than one item and was encouraged to do so. Thus, the percentages do not add up to 100.

oriented families need necessarily sever relations with relatives. Indeed, the evidence is that they tend to see and visit with relatives just as often as parents in troubled families. They may even be exchanging other forms of help. But apparently they are inclined to draw the line on financial dealings. It seems that part of the troubled state of the poorly coping families can be explained by their openness to the easy flow and circulation of their resources. Help is always available to them in some form but at the cost of their being rendered perpetually insolvent.

Concluding Thoughts

This analysis suggests that future-oriented families have chosen a path of striving in poverty that requires discipline and sacrifice in the present for future advantages. They have engaged themselves in a process of *incipient social mobility* to improve their standard of living, to enhance their reputation in their own community, and to raise their children for real socioeconomic mobility. In order successfully to manage scarce resources to achieve these ends, they must devise consumer practices that afford the greatest possible purposeful stretching of every dime and dollar. The rigorous planning and careful expenditures required should be optimal where there is a high level of consensus among spouses or dominance by one and a strong sense of goal-directedness. These kinds of characteristics should, despite stresses and hard times, facilitate pride of accomplishment and family solidarity. But future-oriented families, like all poor families, are extremely vulnerable to environmental forces and events that cannot readily be anticipated or controlled. Their best efforts may come to nought, and they may willy-nilly be forced into the ranks of the troubled poor.

We need not assume that all the troubled families will remain so. The parents in these families were just as well educated and achievement-oriented as those in future-oriented families. Also they were younger (mid-30s versus mid-40s), had a Deep-South, rural, and small-town background, and had not been resident in D.C. as long. In time, many of them may well shift their perspectives on goals and means and gain knowledge and competencies that will enhance their capacity and inclination to cope by using strategies similar to those found in the future-oriented families.

For any of these very poor families, effective coping will not necessarily lead to mobility out of poverty in the parent generation. The structure of power and opportunity in the political economy of the society will play the crucial role in their destiny, not simply their efforts. The complex adaptive processes of black families should not be reduced to simple notions of pathology or health as qualities intrinsic in them. After all, the coping strategies that families choose, often at great cost to themselves, are responses to constraining conditions created and maintained by powerful political and economic sectors of our society. Here is where the real responsibility for family well-being lies, not with those who day in and day out fight the odds in search of a better life or just to survive with self-respect and dignity.

REFERENCES

Aschenbrenner, J. *Lifelines: Black Families in Chicago.* New York: Holt, Rinehart and Winston, 1975.

Bernard, J. *Marriage and Family among Negroes.* Englewood Cliffs, N.J.: Prentice-Hall, 1966.

Billingsley, A. *Black Families in White America.* Englewood Cliffs, N.J.: Prentice-Hall, 1968.

Billingsley, A. Black families and white social science. *Journal of Social Issues,* 1970, *26,* 127-142.

Blassingame, J. W. *The Slave Community.* New York: Oxford University Press, 1972.

Blood, R. O., Jr., and Wolfe, D. N. *Husbands and Wives: The Dynamics of Married Life.* Glencoe, Ill.: Free Press, 1960.

Coleman, J. S. *Resources for Social Change: Race in the U.S.* New York: Wiley-Interscience, 1971.

Etzkowitz, H. and Schaflander, G. M. *Ghetto Crisis: Riots or Reconciliation?* Boston: Little, Brown, 1969.

Farley, R. Trends in racial inequalities: Have the gains of the 1960s disappeared in the 1970s? *American Sociological Review,* 1977, *42* (April), 189-208.

Farley, R. and Hermalin, A. I. Family stability: A comparison of trends between blacks and whites. *American Sociological Review,* 1971, *36* (February), 1-17.

Featherman, D. L., and Hauser, R. M. Changes in the socioeconomic stratification of the races. *American Journal of Sociology,* 1976, *28* (November), 621-651.

Frazier, E. F. *The Negro Family in the U.S.* Chicago: University of Chicago Press, 1939.

Fusfeld, D. R. The Basic Economics of the Urban and Racial Crisis. *Conference Papers for the Union of Radical Political Economics,* December 1968.

Genovese, E. D. *Roll, Jordan, Roll: The World the Slaves Made;* New York: Random House, 1974.

Glick, P. C. A demographer looks at American families. *Journal of Marriage and the Family,* 1975 (February) 15-26.

Glick, P. C., and Mills, K. M. Black Families: Marriage Patterns and Living Arrangements. Paper presented at W. E. B. DuBois Conference on American Blacks, Atlanta, Georgia, October 1974.

Gutman, H. C. *The Black Family in Slavery and Freedom: 1750-1925.* New York: Pantheon Books, 1976.

Hampton, R. Marriage disruption: Some social and economic consequences. In G. J. Duncan and J. N. Morgan (Eds.), Five Thousand American Families: Patterns of Economic Progress, Vol. 3. Ann Arbor, Mich.: Institute for Social Research, 1975, pp. 163-187.

Hannerz, U. *Soulside: Inquiries into Ghetto Culture and Community.* New York: Columbia University Press, 1969.

Heiss, J. On the transmission of marital instability in black families. *American Sociological Review,* 1972, 37 (February), pp. 82-92.

Heiss, J. *The Case of the Black Family: A Sociological Inquiry.* New York: Columbia University Press, 1975.

Hill, R. B. *Strengths of Black Families.* New York: Emerson Hall, 1971.

Hyman, H. H. and Reed, J. S. Black matriarchy reconsidered: Evidence from secondary analysis of sample surveys. *Public Opinion Quarterly,* 1969, 33 (Fall), 346-354.

Jackson, J. J. But where are the men? *The Black Scholar,* 1971, 3 (December), 30-41.

Ladner, J. A. *Tomorrow's Tomorrow: The Black Woman.* New York: Doubleday and Company, 1971.

Lewis, H. *Blackways of Kent.* Chapel Hill: University of North Carolina Press, 1955.

Lewis, H. Culture, class and family life among low-income urban Negroes. In A. M. Ross and H. Hill (Eds.), *Employment, Race and Poverty.* New York: Harcourt, Brace and World, 1967, pp. 149-172.

Liebow, E. *Tally's Corner.* Boston: Little, Brown, 1966.

McWorter, G. The ideology of black social science. *The Black Scholar,* 1969, 1 (December), 28-35.

Morgan, J. N., David, M., Cohen, W., and Brazer, H. *Income and Welfare in the United States.* New York: McGraw-Hill, 1962.

Moynihan, D. P. Employment, income and the ordeal of the Negro family. *Daedalus,* 1965 (Fall), 745-770. (a)

Moynihan, D. P. The Negro Family: The Case for National Action. U.S. Department of Labor, March 1965. (b)

National Urban League. Estimating the 1970 Census Undercount for State and Local Areas. Research Department, Washington, D.C., 1973.

National Urban League. Black Families in the 1974-75 Depression. Research Department, Washington, D.C., 1975.

National Urban League, Quarterly Economic Report on the Black Worker, Report No. 8. Research Department, Washington, D.C., Third Quarter, 1976.

Pettigrew, T. S. *A Profile of the Negro American.* Princeton, N.J.: Van Nostrand, 1964.

Pope, H. and Mueller, C. W. The intergenerational transmission of marital instability: Comparisons by race and sex. *Journal of Social Issues*, 1976, *32* (1), 49-66.

Rainwater, L. Crucible of identity: The Negro lower-class family. *Daedalus*, 1966 (Winter), 72-216.

Rainwater, L. *Behind Ghetto Walls: Black Families in a Federal Slum.* Chicago: Aldine, 1970.

Rainwater, L., and Yancey, W. L. *The Moynihan Report and the Politics of Controversy.* Cambridge, Mass.: MIT Press, 1967.

Rawick, G. P. *From Sundown to Sunup: The Making of the Black Community.* Westport, Conn.: Greenwood, 1972.

Ross, H. L., and Sawhill, I. V. *Time of Transition: The Growth of Families Headed by Women.* Washington, D.C.: The Urban Institute, 1975.

Ryan, W. *Blaming the Victim.* New York: Random House, 1971.

Scanzoni, J. H. *The Black Family in Modern Society.* Boston: Allyn & Bacon, 1971.

Stack, C. *All Our Kin: Strategies for Survival in a Black Community.* New York: Harper & Row, 1974.

Staples, R. *The Black Family: Essays and Studies.* Belmont, Calif.: Wadsworth, 1971. (a)

Staples, R. Towards a sociology of the black family: A theoretical and methodological assessment. *Journal of Marriage and the Family*, 1971 (February), 119-138. (b)

Turner, C. R. Some theoretical and conceptual considerations for black family studies. *Black Lines*, 1972, *2* (Summer), 13-27.

U.S. Bureau of the Census. Census of Population: 1970. General Population Characteristics, Final Report PS(1)-B1, United States Summary, 1970. (a)

U.S. Bureau of the Census. Census of Population: 1970. Subjects Report: Family Composition, Final Report, PC(2)-4A, 1970. (b)

U.S. Bureau of the Census. Current Population Reports. The Social and Economic Status of the Black Population of the United States, Series P-23, No. 54, 1974.

U.S. Bureau of the Census, Statistical Abstract of the United States. 1976 (97th ed.), Washington, D.C., 1976.

3

Practical Considerations

Frederick C. Green

I am pleased with this opportunity to react to such an illuminating and well-done paper as that presented by Professor McQueen.

First of all, I want to thank Professor McQueen for giving me an axe to hold over my wife's head. I didn't realize until today that there was such a shortage of black males. I'm certain that now I can go home and tell her and she will appreciate me a little bit more.

Twenty years ago, Ethel Dunham, at the Old Gallinger Hospital here in Washington, D. C., wrote a book on the care of premature infants. She noted that for clearly defined morphologic reasons there was a significant difference in infant mortality between black males and black females and between black males and white females. Given two babies of the same size, she said, the black male was more at risk, because of physical immaturity based on radiologic bone-age evaluation, than the others. Knowing that the infant mortality rate in the black population, particularly the black poor, is four times that of the majority population, one must won-

Frederick C. Green • George Washington University; Children's Hospital National Medical Center, Washington, D. C.

der whether simple accessibility to optimum prenatal care would solve the problem.

It is refreshing for me to discuss a paper that is a sharp departure from the traditional "deficit model" approach taken by many investigators looking at the black family. Many of the studies I have read in the past were so concerned with defining what is wrong and rushing helter-skelter to correct such perceived deficits that the identification of strengths and the attempt to build on those strengths was completely ignored. Professor McQueen has exhorted us to look to and use these strengths of the black family as we set out to improve the well-being of such families. I think that this is a very important lesson that should be heeded.

Obviously, being a pediatrician, I tend to convert problems to a medical model. I remember that Dr. Sidney Farber of Boston used to tell us constantly about "meddling medicine," ignoring the tremendous untapped recuperative powers of the human body for an intrusive therapeutic modality that may or may not be in the patient's best interest.

Consistent with the medical model, pathology comes about as a result of interaction between the host, the agent, and the environment—for example, the children in our ghettos who have lead poisoning. The host is the poisoned child, the agent is the chip of paint, and the environment is the unsuitable housing in which they live. So it seems to me that if I think of the black family as the host, then it is impossible to draw inferences about the status of that host without looking at the agents and the environment that also impact on that family.

I believe that black families are considerably different from white families in many ways, and I think that the critical reason for that difference is that they have been the subject of severe racist repression for over 300 years. This vector of pathology-racism is critically important. Family instability is not only due to the disturbed structure; it may also be affected by the responsiveness of critical supportive systems such as the health, education, and welfare systems that impinge on their lives. How these systems respond to the black family is as important as the structure of the black family in determining functional stability.

I remember Whitney Young's last book, *Beyond Racism*. He said that black children today are "programmed for failure almost from birth." I think that today it would be fair to say that many of them are programmed for failure from conception because of maternal malnutrition and other noxious influences during intrauterine life.

Incidentally, recently when I was in one of the largest marketplaces in Abidjan, Ivory Coast, in West Africa, I observed a phenomenon that will stay with me for a long time. In this massive marketplace, filled with people, mothers, and numerous children, there was one thing that struck me as odd. I did not hear a child crying. Most of the children were being carried on their mother's back, a very intimate infant-mother contact. This constant reinforcement of the infant-maternal bonding can have only salutary developmental benefit.

Let me say just a little about the previously mentioned pervasive racism, a variable I believe to be a critical enhancer of the impact of poverty. I do believe that the white poor are dealt with differently than the black poor in this country. The black poor are assaulted by both personal and institutional racism. Personal racism is a prior negative judgment of individuals or groups and is held in disregard of the facts that contradict this posture. But of even more importance, I believe, is the institutional racism to which these children and families are subjected. These are policies and practices that restrict, on a racial basis, the choices, rights, mobility, and access to services of families. Perhaps these are the most destructive.

We have heard that for the urban black to survive, he/she must be able to adapt. To what and to whom are they to adapt? Are they to adapt to a majority culture with alien values? Or must they adapt to a culture within their own community that may not always reflect their own values? I am not sure. Further, what is the goal of this adaptation? Are we looking for stability of the black family in an unstable world?

When I look at institutional racism, even when the responsible top-level individuals are far from being "racist," I have to applaud and agree with Ambassador Andrew Young when he said that those who are responsible for institutional policies that perpetuate a form of institutional racism when all they need to do is sign a piece of paper are, in and of themselves, whether or not they agree, either "ethnocentric" or "racist."

I would like to emphasize that black families, as Professor McQueen has pointed out, are impotent to control the factors that impinge on their lives. I'd like to quote from *The New York Times* of September 19, 1976: "In 1975, women held only 11% and Blacks only 4%, of the 22,256 positions on Federal advisory committees and commissions," in spite of the mandate of the Federal Advisory Committee Act that "membership is to be fairly balanced in terms of point of view represented and the functions to be performed." There is no more important advisory committee in existence today than the one that advises the president on national health insurance and on the reimbursement for health services. I look at that 30-person commission, and I would say that it is not really representative of the population at large. This is a glaring example of the impotence of the black family.

Professor McQueen indicates that he has difficulty in considering black families to be "developing" after living in this country for 300 years. I agree! However, you are seeing a generation of black kids now who have never heard of bleaching creams or hair straighteners and other things that are now consigned to the garbage can where they should have been in the past. I believe that the black family has changed markedly in the past two decades. During this time, the civil rights movement and the concepts of "black pride and black power" came into being. The pride of being black was and is particularly important. There is a significant difference, I believe, Professor McQueen, between the groups you identify as the "future-oriented" and the "troubled." The troubled were 10 years younger than the future-oriented. They had lived for a shorter period of time in Washington, D. C. There is a difference between spending your developmental years—the first five years of life—in the 1930s as compared to spending them in the 1940s. I have no empirical data to tell me this, but I do know that there probably will be, in the years to come, a significant difference in the coping capacities and the strategies used by the children who were born in the 1960s.

What role does race pride have in the development of the black family, in assuring the stability of that family? Professor McQueen indicated that in his group of 100 randomly selected poor inner-city families, the study cohort was 57: future-oriented, 34; and troubled, 23. I assume that the other 43 had a welfare index of 1 or greater. But it would be interesting to see what the characteristics of that group could possibly have been. I was a little surprised that many of the troubled families tended to include relatives and nonrelated persons. The extended family was far more intact in that group than in the future-oriented group. You elaborated on that point very well at the end, but I think that those who do not have an extended family to fall back on must have savings. Those who do have an extended family to fall back on perhaps do not feel quite as vulnerable because of the lack of savings. I know that in the black community there used to be—and still are—sick clubs, burial clubs, clubs in which groups of people gathered together and pooled their money because there was difficulty in getting the various forms of insurance. This was a "kith-extended care network," a kith network rather than a kin network.

The quest for respectability—the Church versus the tavern—is very understandable. Those in my generation who grew up in the black communities of smaller towns understand the importance and centrality of the Church in such communities. It did not necessarily reflect hyperreligiosity. We couldn't go to the movies because we were discriminated against. We couldn't use the YW or YMCA because they were out-of-bounds. All of the other social institutions of the community were closed to black people. There was one institution that was open: the Church. And this was what we would use. I think it is no accident that some of the great strides that have been made within the civil rights movement and in the black communities of this country have been catalyzed through the Church.

I thank Professor McQueen for such an excellent paper.

Practical Considerations

T. Berry Brazelton

I found Professor McQueen's paper very rewarding. It really focused in on some goal-oriented behavior and goal-oriented processes, and it showed us a group of people represented by a very exciting and important movement: the black movement. This group is identifying strengths, defending them, protecting them, and building them up in the face of so many identifiable pressures.

He presented a coping model of people under cultural stress, and he gave us an opportunity to adapt our pathologically oriented viewpoints to see the coping systems—the coping strengths that are inherent not only in survival but that have contributed to the goal orientation of black people and of the black movement.

Our pathological eyes have dominated our look at black systems for too long. I think the work of many of the writers in our country have been subtly putting down black people. Their works demonstrate how defensive we are about our present white, middle-class standards. I just wonder, as a member of the white middle class, what are we afraid of? Do we indeed have such a

T. Berry Brazelton • Harvard Medical School, Cambridge, Massachusetts.

shaky system and such shaky standards that we daren't allow for individual differences and variability—even treasure them?

The other point that came through to me as I read Professor McQueen's discussion of Moynihan's book is the importance we have placed on reinforcing weaknesses—on keeping subcultures down with a kind of inflexibility. Instead of that, Professor McQueen gives us an opportunity to look at the goals and strengths that the blacks have shown and have preserved over the past 200 years. He points out that in the face of the sort of confusion that all of us feel about what's happening to the family in our country, we can look to their strengths with a real identification with their struggle.

The inflexibility in our middle-class standards that cannot allow for strong differences in blacks or other subcultures is disturbing. Are we so insecure, so stressed in our white middle-class cultural values that we can't allow for and even encourage differences from other subcultures? The young people in our midst are questioning many of our more rigid values and practices and are searching for new ones. Young black families are looking for a heritage, for the family patterns and coping styles that are represented in Professor McQueen's paper. We can all profit by them, for they are indeed an example of the strengths of a people who can survive in an unwelcoming land under many inner and outer stresses.

The most serious threat to the development of a large segment of the black population is that of poverty and the accompanying undernutrition. Not only is it a constant drain on one's emotional development to grow up in poverty, but a more serious drain is that with which a fetus and a small infant is presented when his mother is inadequately nourished in pregnancy. From the work with animals, Winick and others have shown that cellular growth and replication are interfered with when the developing fetal brain is undernourished. The number of cells, their size, and the complexity of their connections with other parts of the brain can all be decreased under such conditions. And, of course, this kind of underdevelopment is likely to affect the newborn's behavioral responsiveness and may well affect his ultimate potential for cognitive and affective development. I've just finished an analysis of data from an extensive study in Guatemala that we did along with

Incap. This study stemmed from my work with neonates. By observing behavior during the neonatal period, we were able to identify, with a 90% accurate prediction, babies who would end up
with kwashiorkor a year and a half later. As you know, kwashiorkor is a nutritional deficiency disease that occurs in the second year
as babies are weaned from breast milk.

A very quiet nonsoliciting baby will get only what he asks for
from an overstressed poverty-stricken environment. When asked
how often they fed their newborns, mothers of babies we'd identified would say, "when he wants it." If you stayed in the household with them, they were fed three times a day in the first few
months after delivery. These babies' quiet underdemanding behavior at birth correlated with their lengths (their linear growth in
the uterus), with their head sizes (potentially correlated with the
number of brain cells they were equipped with at birth), and with
their mothers' height and their mothers' own head size. This correlation in turn predicted their ability to function on standard intellectual assessments in later infancy and childhood. This last correlation has to do with their mothers' own nutritional experience and
the cellular or DNA replication in their own mothers' uteri. Infant
behavior at birth also correlated with the number of days the
mother had to remain in bed around the time of a delivery in a
society that doesn't go to bed at all unless they can't cope. This
might be reflecting the stress on an inadequate maternal system.
Finally, infant behavior was also found to correlate with the
number of hypoxic episodes that these neonates had around delivery in a culture that delivered at home by midwifery, without drugs
or medication. Hence we saw small, quiet, nondemanding infants
whose delivery was already *more* stressful, and whose size and
brain development was at risk before delivery. We were already
talking about a compounded situation, *at birth*, for babies who
were undernourished in the uterus, whose brain cells were potentially endangered by hypoxic episodes, and who faced life in a culture that wasn't going to nurture them properly because of their
unresponsive neonatal behavior.

To talk about poverty as if it were a one-generational affair is
no longer feasible. It's at least a three-generation affair. We cannot
fault the poor for not getting jobs, for standing around on street

corners, and for not being able to learn in school, when the number of cells, the DNA in their brains is cut down by perhaps 40%, when their thyroids, their adrenals, all the primary organs that contribute to activation, to motivation, to prolonged attention—all necessary skills for survival in a complex culture like ours—are interfered with.

Another rather specific effect that I think our culture has on the black self-image is something I would like to pose more as a question. It's something that I personally have been questioning a lot because I'm so interested in early mother-infant interaction and what it means to the infant and to his self-image. I've been aware, and struck clinically, by the fact that black babies at birth are very exciting to play with, that their motor development at birth is a real asset, and that they are ahead of Caucasian babies in motor responsiveness. Geber first described this phenomenon in babies in Africa. She identified it as a kind of motor "precocity" at birth, as compared to European babies. I don't like the idea of precocity: I think precocity sounds as if it were fixed and were an expensive notion and one that, in this country at least, gets equated with an energy model. Jensen, Shockley, and various others who have put blacks down have indicated that if there were motor precocity, there must be a cognitive deficit for black children. I think that these comments have led black mothers to believe that if a baby is exciting in motor development, he will be underdeveloped in the cognitive and affective spheres. I find those concepts extremely destructive for a young mother trying to establish a relationship with a child. For she puts down her natural responses to his motor behavior and tries to curb them in him, and he gets the feeling that he is "no good."

What happens with this motor excitement in infants in Kenya, where we've been working, is that mothers pick up on it and begin to reinforce it. They play with babies, actively entering into an easy relationship built around the motor excitement of carrying and playing with their baby with a lot of motoric interaction. Of course their babies are *not* delayed in affective or cognitive development as a result of this kind of interactive reinforcement.

In this country, if you tell a black mother her baby is exciting to play with and is ahead in his motor development, she may well say, "Yeah, he's fresh" or "He's bad." She'll say it with a smile, but at the same time, I think she's reflecting a cultural put-down for that kind of behavior that invades her thinking about her baby at some level. And I think our white middle-class culture has been reinforcing this kind of thinking.

I've found in my practice that many black babies do cry more and are harder to keep quiet than are white middle-class babies in the first few months. Mothers call up and say, "I can't manage him." I say, "Why don't you carry him around?" They say, "Oh, but he's so busy, I want to quiet him down," and I say, "But why?" And they say, "If I don't quiet him down now, he'll never make it when he has to sit down in first grade."

Now this occurs in the first month or two of life; parents are already equating neonatal behavior and the excitement that might be present in such a motor sphere for the baby with what's going to be construed as a put-down by society in first grade. My feeling is that we'd better change first grades, and we certainly should begin to reinforce young mothers for the individual differences they find in their new infants and for the excitement that's available to them in whatever area they present. It seems to me that when a mother continually puts down the child's capacity to respond, perhaps because she is feeling repressed herself, she is passing on a poor self-image to that baby at a time when he can't fight back. That self-image is bound to affect him later. I would like to suggest that we as pediatricians and others working with new young parents in the area of early intervention and prevention can change that put-down for young parents and for their babies. I think the most exciting thing about Professor McQueen's paper is that forward-looking goal-oriented families in the black segment haven't given up yet.

REFERENCES

Brazelton, T. B., Tronick, E., Lechtig, A., and Lasky, R. The behavior of nutritionally deprived Guatemalan infants. *Developmental Medicine and Child Neurology*, 1977, *19*, 364-367.

Geber, M., and Dean, R. F. A. The state of development of newborn African children. *Lancet*, 1957, *2*, 1216.

Winick, M. Cellular changes during placental and fetal growth. *American Journal of Obstetrics and Gynecology*, 1971, *109*, 166.

III

Child Development and New Family Forms

4

The Development of Children in Mother-Headed Families

E. Mavis Hetherington, Martha Cox, and Roger Cox

The incidence of children growing up in a mother-headed single-parent home in the United States is increasing. Over 10 million children currently are living in fatherless homes, largely because of the rapidly accelerating divorce rate, particularly in families that have young children. Fewer parents are "staying together for the sake of the children." In 1975, there were more than 1 million divorces, and in each divorce an average of 1.22 children were involved. In contrast to the increasing divorce rate, the marriage rate has begun to drop. The ratio of divorces to marriages in the United States in 1975 was about 1:2, with 1,026,000 divorces and 2,126,000 marriages.

In divorces in which children are involved, except under unusual circumstances, the mother gains custody of the child. Al-

E. Mavis Hetherington • University of Virginia, Charlottesville, Virginia. Martha Cox • University of Evansville, Evansville, Indiana. Roger Cox • Indiana State University, Terre Haute, Indiana.

though the proportion of children living with their divorced fathers is increasing, in the 1975 Population Survey by the U. S. Census Bureau, only 8.4% of children in single-parent homes were reported to reside with their father. Thus, the most frequently found family condition in the immediate post-divorce situation is one in which a child is living in a home with a single mother and is having intermittent or no contact with the father.

There is a large body of research literature dealing with the development of children raised in mother-headed single-parent families. In many of these studies the one-parent family is approached with the assumption that it is not only a variant of the nuclear family but also a pathogenic variant. In addition, these studies customarily have been labeled *father-absence* studies. This rubric reflects a bias that when differences are found between children in single-parent and in intact families, they are attributable to the absence of the father rather than to differences in family functioning, stresses, and support systems in the two types of families.

In this paper, we will focus on the functioning of the mother and the child who are present in the home rather than on the absent father, and on family process rather than on family structure. Some of the research findings relating to the development of children in single-parent mother-headed families will be reviewed. An attempt will be made to identify social factors and characteristics in family interaction and functioning that may contribute to the social and cognitive development of these children. No attempt will be made to review the literature on extreme forms of psychopathology.

STRESSES IN SINGLE-PARENT FAMILIES

How does the life-style and functioning of mothers and children in single-parent families differ from those of nuclear families? What stresses and patterns of coping and adjustment are most likely to be encountered in single-parent families?

First, mothers in single-parent families suffer from task overload. Glasser and Navarre (1965) have discussed cogently the problems confronting the single parent in dealing with family tasks and needs that are regarded as a full-time job for two adults in a nuclear family. Although household tasks are not intrinsically sex-differentiated, many women in single-parent families, especially women from marriages in which traditional sex-stereotyped roles have been maintained, have particular difficulty in dealing with economic problems, household maintainance, and tradespeople. If adequate support systems are not available, something has to give. This has been reported to result in a higher level of family disorganization in single-parent families where the household routines and roles are less systematically assigned and performed (Hetherington, Cox, and Cox, 1977).

Second, single-parent mother-headed families often experience financial duress. In 1974, the mean family income of mother-headed families was $6,413 in contrast to a mean income of $13,788 for male-headed families. Female-headed families with children experience more economic deprivation than those without children. In 1974, 51.5% of children under 18 in female-headed families were in families with incomes below the poverty level (Bane, 1977).

Such economic hardship frequently forces women to work who would otherwise prefer to remain in the home or who are not highly enough skilled to obtain positions that can pay for adequate child care. There is evidence that with satisfactory child supervision, maternal employment is not deleterious to children and may have positive effects on the sense of competence and on the maintenance of social contacts of mothers in single-parent families. However, if the quality of substitute child care is inadequate or if the mother begins to work around the time of dissolution of the marriage, so that the child experiences the double loss of father and mother simultaneously, maternal employment is associated with a high rate of behavior disorders in children (Hetherington *et al.*, 1977).

Third, mothers in single-parent families are often socially isolated and lacking in social and emotional support. Although divorced women maintain ties with their own kin, those with their husband's kin are minimal, and they are likely to have fewer friends, belong to fewer organizations, and participate in fewer recreational activities than do married women (Anspach, 1976; Hetherington *et al.*, 1977; Marsden, 1969; Pearlin and Johnson, 1975; Spicer and Hampe, 1975; Stack, 1972). This social isolation is often associated with acute depression (Hetherington *et al.*, 1977). One of the findings that is notable is that family contacts with parents and siblings do not ameliorate this loneliness. A social network and the formation of friendships and intimate attachments seem critical (Pearlin and Johnson, 1975; Hetherington *et al.*, 1977).

It might be thought that the presence of children would attenuate this sense of loneliness; however, recent studies (Hetherington *et al.*, 1977; Weiss, 1977) suggest that the presence of children may actually make mothers feel more unhappy, frustrated, helpless, anxious, and incompetent. This was particularly true of divorced mothers with noncompliant acting-out young sons. These mothers often complained of being prisoners, of being trapped, or of being locked in a child's world. One mother stated it cogently when she said, "There's no time-outs in the parenting game in one-parent families."

Even the most loving involved parent has times when his or her parenting benefits from the opportunity for a little distancing from a recalcitrant or demanding child. Another parent can provide the time-outs that ease some of the stresses of parenthood.

This leads to the final way in which single-parent families differ from nuclear families. In a single-parent family, there is only one parental figure to serve as the agent of socialization through discipline, by direct tuition, or by acting as a model. Thus, the remaining parent is likely to become more salient in the development of the child. There is not a spouse to serve as a buffer between parent and child in a single-parent family. In nuclear families, a loving, competent, or well-adjusted parent can help counteract the effects of a rejecting, incompetent, emotionally unstable parent.

The single mother may confront specific problems of authority in discipline. Children view fathers as more powerful and threatening than mothers. In nuclear families, children exhibit less noncompliant and deviant behavior toward their fathers than toward their mothers, and when undesirable behavior occurs, the father can terminate it more readily than the mother can (Hetherington *et al.*, 1977). The single mother may have to be supermother to counter the image of greater authority and power vested in males in our culture.

The single parent also offers the child a more restricted array of positive characteristics to model. A mother and a father together are likely to exhibit wider-ranging interests, skills, and attributes than a single parent.

This is not meant to be a panegyric to the nuclear family. The rising divorce rate is just one manifestation of dissension in the nuclear family. Research findings are consistent in showing that children reared in conflict-ridden nuclear families are more poorly adjusted than children in well-functioning single-parent homes (Nye, 1957; McCord, McCord, and Thurber, 1962). In nuclear families, anti-social behavior in children increases as the quality of the marital relationship deteriorates (Rutter, 1971).

In addition, in many legally intact families, a psychological and emotional divorce has occurred. Many husbands are detached and inaccessible to their wives and children. As will be seen, these psychologically unavailable fathers are little better than a physically absent father. In such alienated or conflicted families, separation may be the most beneficial outcome to an emotionally destructive relationship.

The point that must be emphasized is that in single-parent mother-headed families, the mother is confronted by many stresses that threaten her social and psychological well-being. Since she assumes the pivotal salient parental role, the way in which she copes and adapts to these stresses may be determinative in shaping the development of her children.

METHODOLOGICAL CONSIDERATIONS

The methodological considerations in studies of single-parent families are complex. The reason for the separation of the parents, the length of the separation, and the age of the child at the onset of separation have all been found to effect some aspects of the development of the child. Also, specific characteristics of the child such as sex, age at testing, race, socioeconomic status, and birth order influence the impact of separation and growing up in a single-parent home. No studies have adequately controlled or assessed the influence of all of these variables. In addition, since much of this research has been conceptualized as the study of the impact of father absence, most of the studies have used male subjects, since fathers often have been regarded as playing a more important role in the development of boys than of girls.

Research on single-parent families encompasses an enormous number of studies; therefore, no attempt will be made to present a complete review of the literature. Instead, inferences and speculations will be drawn on the basis of methodologically superior studies, and the findings of a few selected studies will be presented in greater detail for illustrative purposes. This paper will focus on the intellectual and social and emotional development of children in single-parent families.

INTELLECTUAL DEVELOPMENT

Studies of the intellectual development of children from single-parent families have investigated both the overall level of intellectual performance and achievement and the patterning of intellectual abilities. Two recent reviews of this research have concluded that children growing up in mother-headed families show deficits in cognitive performance as assessed by standardized intelligence and achievement tests and school performance (Biller, 1974; Shinn, 1977). In her survey of this literature, Shinn (1977) reported that the differences in cognitive performance between children from nuclear and single-parent families is considerable: "from 0.2 to 1.6 years in achievement, from 0.2 to 0.9 standard deviation units in I. Q. and aptitude, and 0.8 of the difference between and 'B' and 'C' in grade point average" (p. 5).

What factors may be related to these differences? Since single-parent families are more frequent among black and economically deprived children and since such children perform less well·than middle-class and white children on tests of intelligence and achievement, it is essential to control for social class and race in evaluating differences in cognitive performance. However, even when appropriate controls are instituted, these differences remain, although as will be seen, the pattern of test scores may differ for lower- and middle-class children.

In general, although the effects are most marked for boys, they are found in both sexes. Birth order rarely has been considered in these studies; however, one investigator has found that first-born children are more cognitively disadvantaged than are later borns in single-parent families (Cox, 1974).

In addition, some studies find that early loss of the father is more deleterious than later separation and that the effects may be late appearing and cumulative. Differences in the intellectual performance of children from single-parent and nuclear homes are rarely found in the preschool years; they emerge and increase over the course of development in the school years (Deutsch and Brown, 1964). Several longitudinal studies, where repeated measures over time on the same children were available, found no deficits when the children were young but the emergence of deficits with increasing age (Hess, Shipman, Brophy, and Bear, 1968; Hess, Shipman, Brophy, Bear and Adelberger, 1969; Hetherington, Cox, and Cox, 1978; Rees and Palmer, 1970). It may be that the types of items on preschool tests that involve rote memory, simple vocabulary, and sensorimotor skills are less sensitive to factors associated with being raised in a single-parent family than the problem-solving and abstract-reasoning items found on tests for older children.

Variations in the availability of fathers occur not only between mother-headed and nuclear families but within nuclear families. Some fathers, because of occupational demands or rejection and lack of involvement with the family, are not available to their children. Children in nuclear families with unavailable fathers show decrements in achievement similar to those in mother-headed families. Blanchard and Biller (1971) studied the effects of early (before age 5) versus late father absence and low father availability (less than six hours a week) versus high father availability in nuclear homes with third-grade boys. They found that boys with highly available fathers surpassed the other three groups in achievement test scores and classroom grades. The early father-absent boys were found to be underachievers. The boys from nuclear families with relatively unavailable fathers and the late father-absent boys were also found to be below grade-level expectations, although not as severely as boys who had been separated from their fathers before the age of 5. Other studies have found that children whose fathers have been temporarily absent for two years or less and children whose fathers were in the military or worked on nightshifts and presumably were less available to their children have lower scores on achievement tests. These deficits are found even in young adulthood on the American College Entrance Examination.

The father with low participation in the family or with low warmth in his relationship with his family may be just as detrimental to the child's intellectual growth as one who is totally absent. This may in part account for the frequently reported finding of greater negative effects of divorce when compared to death of the father. It seems possible that in many cases the divorce may have been preceded by psychological or actual withdrawal of the father from the family setting. Clearly, in nuclear families, the presence of the father is not the important variable; the important variable is the participation of a good father.

The point has been made that in some cases, father absence caused by divorce may in fact enhance the child's development (Lamb, 1977). If the child's loss of an uninterested father results in the acquisition of or association with a more accessible and suitable father figure, the child may benefit from the results of this transition. In support of this theory, several investigators have reported that the cognitive performance of children living with father surrogates such as stepfathers is superior to that or is not different than that of children in other nuclear families and is superior to the performance of those in single-parent families (Lessing, Zagorin, and Nelson, 1970; Santrock, 1972; Solomon, Hirsch, Scheinfeld, and Jackson, 1972).

In addition to studies of the overall level of cognitive performance, many investigators have studied the patterning of cognitive abilities. Aptitude and achievement tests in the general population have shown that females are usually superior to males in verbal areas, whereas males are superior to females on quantitative tasks. In single-parent families, both male and female children are more likely to show the female pattern of higher verbal than quantitative scores. In most cases, this finding is associated with a decrement in quantitative scores; however, in the middle-class relatively elite samples such as college students from Harvard, Stanford, or Carleton or in children from upper-middle-class families, being raised in a single-parent family may actually enhance verbal scores (Carlsmith, 1964; Funkenstein, 1963; Oshman, 1975; Lessing *et al.,* 1970). This enhancement has never been found in children from lower socioeconomic levels.

Several explanations have been advanced to account for these differences in cognitive development. The first is that fathers are more likely than mothers to present an independent, achievement-oriented, problem-solving, quantitatively skilled model to their children. In the absence of the father, children identify with the more verbally fluent but less achievement-oriented mother. In this case, it might be expected that if the mother offers the child a high-achieving model and rewards the child for self-directed task-oriented behavior, the effects of absence of a male paternal model would be mitigated.

The second explanation is that considerable anxiety-provoking stress occurs in single-parent families and that anxiety is more likely to interfere with problem-solving skills than with verbal skills (Nelson and Maccoby, 1966). Thus a mother-headed family would be only one stressful family situation in which a depression on quantitative scores as a result of anxiety might occur. In support of this position, both loss of a father and a mother and high conflict in nuclear families are related to low quantitative scores. However, several studies have indicated that divorce, which one might assume is likely to be preceded by considerable conflict, has more negative effects in depressing overall cognitive performance—and not just on quantitative performance—than do other types of separation (Crescimbeni, 1965; Santrock, 1972; Sutton-Smith, Rosenberg, and Landy, 1968). Adverse intellectual effects on overall scores have even been obtained immediately preceding separation due to divorce, desertion, or separation (Santrock, 1972). Thus, it is apparent that stress and anxiety have an adverse effect on cognitive performance, but how these relate to differential patterning of abilities in single-parent families is not clear.

The third proposal is that children from single-parent homes receive less adult attention than do children in nuclear families and that the amount of interaction with parents is a crucial factor in intellectual achievement (Shinn, 1977; Zajonc, 1976).

A recent longitudinal investigation that studied stress, coping, and adaptation in divorced parents and their middle-class, white preschool children in the two years following divorce supports this position (Hetherington et al., 1977). These were divorced families in which the father continued to have some contact with his child. At two months after divorce, one quarter of the fathers were actually seeing more of their children than they had during the marriage; however, over the two-year period, the fathers became increasingly unavailable to their children. In addition, disorganization in routines and maternal employment in divorced families led to many of the divorced mothers, becoming less available to their children. Divorced mothers were less likely to read to their children at bedtime, prolong child-care routines in a playful way, or eat with their children. In this study, the availability of both parents was correlated with the child's performance on standardized cognitive tasks. However, the influence of the father on both the intellectual and the social-emotional development of the child declined over time, whereas the divorced mother became increasingly salient in her child's development.

In some homes, single mothers, especially single middle-class mothers may attempt to compensate for the loss of the father. Shinn (1978) proposed that

> single mothers who are coping with adverse economic circumstances may have little left over for their children, while single mothers who are financially better off may spend additional time with their children which they would otherwise have devoted to their husband.

She went on to say that with a highly verbal middle-class mother, these interactions are likely to be of the kind to promote verbal development. This could explain the enhancement effects of verbal test scores in mother-headed families that were found only in relatively elite populations and never in economically deprived samples.

A final interpretation can also be proffered. In the longitudinal study by Hetherington et al. (1977), it was found that in divorced families there was a marked breakdown of appropriate and consistent parental control over children; fewer demands for mature independent behavior; and less communication, explanation, and reasoning with children. These poor parenting practices were associated with high distractibility, impulsivity, short attention spans, and lack of persistence on tasks by the children, which in turn were associated with drops in scores on performance and quantitative tasks and on certain types of problem-solving tasks. Problem-solving and academic success require the ability to concentrate and persist. This ability to focus and sustain attention seems more critical in tasks that involve reasoning such as mathematical problem-solving than on such things as vocabulary, hence, the frequently reported quantitative-verbal discrepancy found in children in mother-headed families. Children of divorced mothers who were available, maintained firm but sensitive discipline, communicated with their children, and encouraged independent mature behavior showed no cognitive deficits. What is being proposed is that poor parental control leads to high distractibility and lack of persistence in children, which causes poor problem-solving performance. It would seem that the quality as well as the quantity of maternal interaction in single-parent families should be considered.

SOCIAL AND EMOTIONAL DEVELOPMENT

Sex-Role Typing and Heterosexual Relations

The most extensively studied aspects of social development in single-parent families are the development of sex-role typing and of self-control. Sex-role typing is the process by which children acquire the motives, values, and behaviors regarded as characteristically masculine or feminine in their culture. Although changes are occurring in sex-role standards, particularly in the beliefs of middle-class women, most adults and children still maintain rather stereotyped traditional conceptions of sex roles. Males in our society are viewed as more assertive, independent, dominant, and competent in dealing with problems in the environment. In contrast, females are thought to be more submissive, nurturant, emotional, and sensitive in interpersonal situations.

The commonly held view has been that children learn sex-typed behaviors by identifying with or imitating the same-sexed parent. Therefore, studies of father absence have focused on the sex-role development of boys. However, research findings do not substantiate this position. Fathers seem to play a more important role than mothers in the sex typing not only of boys but also of girls in nuclear families. Boys reared by warm dominant masculine fathers are themselves more masculine in identification, in overt behaviors, and in stated preferences for various sex-typed activities and roles (Biller and Borstelmann, 1967; Hetherington, 1967). In turn, masculinity, warmth, involvement in child rearing, and high evaluation of and reinforcement for feminine behaviors by fathers is associated with feminine development in girls (Hetherington, 1967; Mussen and Rutherford, 1963; Sears, Rau, and Alpert, 1965). Surprisingly, reinforcement for sex-typed activities and femininity in the mother are not related to femininity in daughters. This may in part be because fathers are much more concerned than mothers about the maintenance of stereotyped sex-role behavior in their children and are more likely to vary their role as they relate to male and female offspring. Depending on your view of stereotypical sex roles, the father can be regarded either as an effective parent or as the villain in the piece. In nuclear families, the father's rigid view of sex roles may actually interfere with the development of independent achievement-oriented behaviors in girls and with the interpersonal sensitivity and the freedom to express tender emotions in boys.

On the basis of these findings in nuclear families, it might be expected that deviations in sex-role typing would occur in both boys and girls when the father is absent; however, this is not the case. Boys but not girls from mother-headed families show some disruptions in sex typing but only if separation from the father has occurred before the age of 5. There are few differences in the sex-role typing of boys separated from their fathers after the age of 5 from that of boys in nuclear families. This is probably because sex-role typing in most children is largely established by age 5. Preschool boys and boys in their early school years from mother-headed families have been described as more dependent, less masculine (Santrock, 1970), more feminine in self-concepts, and less aggressive than boys from nuclear families. They may show play patterns and game preferences more characteristic of girls than of boys. They also have been reported to show greater verbal than physical aggression, which is a pattern more commonly found in girls (Bach, 1946; Sear, 1951).

Although preschool boys from mother-headed homes exhibit disruptions in sex-role typing, the results for adolescent boys are less consistent. Many studies of older boys report no differences in the sex-typed behavior and sex-role preferences of boys from nuclear and single-parent homes (Barclay and Cusumano, 1967; Biller and Bahm, 1971). This is especially true in studies of boys of lower socioeconomic status. Other studies report that compensatory masculinity may occur. Compensatory masculinity involves inconsistent patterns of extremes in sex-typed behavior. Boys exhibiting compensatory masculinity may at times manifest excessively masculine, assertive forms of behavior and at other times show feminine behaviors such as dependency. It has sometimes been argued that such compensation is a result of the father-absent boy's desperate attempts to maintain a masculine identification when no masculine role model is present. In such a situation, rather than acquiring a stable masculine identification, the boy acquires a set of loosely integrated responses that appear to be almost a caricature of the stereotyped masculine role. It has been found that boys raised in mother-headed families are more likely than boys from nuclear families to show a pattern of aggressive behavior plus either high dependency on adults or homosexual tendencies (McCord et al., 1962).

Miller (1958) proposed that the toughness and hypermasculine behavior often found among lower-class adolescent male gangs may be due to the increased proportion of fatherless homes in the lower classes. This exaggerated masculinity, which may be expressed in the form of delinquent behavior, reflects an attempt to compensate for the feminine orientation they have derived from their home life.

If compensatory masculinity occurs, it may in part be attributable to Biller's (1974) finding that the unconscious identification of boys from mother-headed families remains feminine long after social pressures shape overt behavior and preferences for the male role in a masculine direction.

It might be speculated that with increasing age, most boys become aware of the greater privileges and status of males in our society and develop a preference for the masculine role. However, the facility with which they are able to acquire and perform appropriately masculine behaviors will depend to a large extent upon the availability of masculine models in siblings, stepfathers, grandfathers, teachers, peers and the mass media. The presence of an older male sibling is associated with greater masculinity in boys in both mother-headed and nuclear families (Santrock, 1970; Wohlford, Santrock, Berger, and Liberman, 1971).

We have been arguing that the functioning of the mother who is present may play a more important role than simple father absence in the development of children in mother-headed single-parent families. Although in nuclear families the mother does not play the key role in shaping the sex typing of her children, she may be more critical in a single-parent family. There is some evidence that mothers in single-parent homes may attempt to be more restrictive, and overprotective and less encouraging of masculine activities than are mothers in nuclear families (Biller and Bahm, 1971; Hetherington et al., 1977). In the previously cited longitudinal study of the aftermath of divorce by Hetherington et al. (1978), in single-parent families but not in nuclear families these maternal variables plus her attitude toward her husband related to the sex-role typing of sons. When divorced mothers encouraged independent mature positive masculine behaviors and had a positive attitude toward males and their ex-husbands, boys in divorced families did not differ from those in nuclear families on measures of sex-role typing. In addition, such mothers who also exhibited and valued interpersonal sensitivity and nurturance had children who were highly androgynous, showing positive attributes associated with both masculinity and femininity. By two years after divorce, few correlates were found between the sex-role typing of children and their father's attributes or behaviors. Again this suggests that whereas certain maternal behaviors may not play a role in shaping children's development in nuclear families, they become potent in the single-parent mother-headed family. We cannot generalize to the impact and the mediating role of maternal behaviors in single-parent families from findings based on nuclear families.

There have been few studies of sex-role typing in girls from single-parent families. Those that have been done tend to find no disruptions in traditional measures of sex-role typing in young girls from mother-headed families (Bach, 1946; Lynn and Sawrey, 1959; Santrock, 1972). Differences do emerge in adolescence and young adulthood, but these seem to be restricted to the area of heterosexual relationships. It could be argued that one of the most important ways in which sex typing is manifested is in satisfactory relations with members of the opposite sex.

Early interactions of daughters with a loving attentive father may be an effective and nonstressful way of acquiring the attitudes, skills, and confidence that facilitate successful heterosexual relations. In nuclear families, women who have had close warm relationships with their fathers are more likely than women with aloof unaffectionate fathers to have happy marriages and a high rate of orgasmic satisfaction (Biller, 1976; Fisher, 1973; Johnson, 1975). Inadequate alienated fathers also were frequently found in the families of both male and female homosexuals (Apperson and McAdoo, 1968; Bene, 1965; Bieber, 1962; Evans, 1969; Saghir and Robbins, 1973; Thompson, Schwartz, McCandless, and Edwards, 1973).

Women from mother-headed families or families in which there was frequent early separation from the father are more likely to have unsatisfactory sexual relationships (Fisher, 1973; Jacobson and Ryder, 1969). In addition, there seems to be a generational transmission effect of marital instability, and both the male and the female children from single parents are themselves more likely to divorce. This does not seem to be attributable solely to having only one parent, since it is more likely to occur if the child has lived with a single mother than a single father and in families where separation has been caused by divorce rather than by death. It may be that greater stresses and different life-styles are associated with being a single mother than with being a single father and with being divorced than with being widowed (Pope and Mueller, 1976; Mueller and Pope, 1977).

One study (Hetherington, 1972) has examined in detail some of the differences in heterosexual behavior of adolescent, white, lower- and lower-middle-class girls from divorced, widowed, or nuclear families. It was found that girls from the three types of families exhibited very different patterns of behavior in relating to male peers or adults.

A disruption in relationships with males in girls from mother-headed homes appeared either as excessive sexual anxiety, shyness, and discomfort around males or as sexually precocious and inappropriately assertive behavior with male peers and adults. The former syndrome was more common when separation had occurred because of the father's death, the latter when separation was result of divorce. These behaviors did not occur in interacting with females. The behavior of these girls was observed in a number of settings including a recreation center they attended. In recreation center dances, males tended to congregate at one end of the hall and females at the other. The behavior of the girls was recorded on these occasions. The first group, inhibited daughters of widows, tended to remain in the cluster of other girls unless they were invited to dance. They more frequently positioned themselves in the back row than did the other group of father-absent or father-present girls. Some even spent over 90% of the evening hiding in the ladies' room. In contrast, the second group, the daughters of divorcées, spent more time at the boys' end of the hall, more frequently initiated encounters and asked male peers to dance, and more frequently touched the males in proximity to them. This was not related to differences in popularity, since girls in nuclear families and both separated groups were asked to dance equally often when they were in the hall.

Except for behaviors associated with proximity and attention seeking or avoidance of males, there were few differences between girls with absent fathers or present fathers in traditional sex-typed behaviors, activities, preferences, or interests.

The girls were also observed in the laboratory while being interviewed by a male or a female interviewer. Daughters of divorcées sat closer to the male interviewer and showed greater openness of body position and greater eye contact and body orientation toward the interviewer than did the daughters of widows. Daughters of widows showed little eye contact or smiling and also maintained a rigid closed body position oriented away from the interviewer. The behaviors of girls from nuclear families fell between those in the two single-parent groups. There were no differences in the way the three groups of girls responded to the female interviewer. Therefore, these responses were specific to interactions with males.

Again let us look at the behavior of the mothers in attempting to understand these differences. There was greater conflict, inconsistency, punitiveness, and restrictiveness about sex by divorced mothers than by mothers in the other groups; however, this might well be a response to their daughters' acting-out sexual behavior.

It was mainly in attitudes toward herself, her marriage, and her life that the divorcée differed from the widow. The divorcée was anxious and unhappy. Her attitude toward her spouse was hostile, and her memories of her marriage and life were negative. These attitudes were reflected in the critical attitude of her daughter toward the divorced father. Although the divorcée loved her daughter, she felt that she had had little support from other people during her divorce and times of stress and with her difficulties in rearing a child alone. This is in marked contrast to the positive attitudes of the widows toward marriage, their lost husbands, the emotional support of friends and family at the loss of a husband, and the gratification of having children. These attitudes were reflected in the happy memories their daughters had of their fathers.

Both groups of girls from single-parent families reported feeling anxious around males but had apparently developed different ways of coping with this anxiety. It may be that daughters of divorcées viewed their mothers' separated lives as unsatisfying and felt that for happiness it was essential to secure a man. It may also be that life with a dissatisfied anxious mother—even if she loves the daughter—is difficult, and that these daughters were more eager to leave home than daughters of widows living with relatively happy secure mothers with support from the extended family. In contrast, daughters of widows with their aggrandized image of the father may have felt that few other males could compare favorably with him or alternately may regard all males as superior and as objects of deference and apprehension. One can also speculate that after experiencing the sudden loss of a father, they were unwilling to make another deep emotional commitment to a male.

A follow-up study was conducted on the girls in the study to see how long-lasting the effects of father absence on daughters' relationships with men were, as reflected in the girls' subsequent marital relationships. Daughters of divorcées not only married younger, but more of them were pregnant at the time of marriage and several are already separated or divorced. No differences in frequency of marital intercourse were reported for the three groups, but orgasmic satisfaction was less in girls from single-parent families.

Freud has suggested that girls continue to relive their relationships with their fathers through subsequent interactions with men. Do girls marry men who resemble their image of their fathers?

On an adjective checklist on which the girls checked adjectives that were most or least like their husband, father, or most men, daughters of both divorcées and widows reported more similarity between their husbands and fathers than did girls from nuclear families. Freud may be right, but only for father-absent girls. Girls from mother-headed families may not have the opportunity to work through their feelings about their fathers as do girls with the father present, and they may seek to resolve these feelings with their mates. Girls who have had a continuous relationship with a father may come to view their husbands in a more balanced realistic way, perceiving virtues and shortcomings. Girls with absent fathers maintain their childhood image of their fathers. Daughters of widows perceived both their husbands and their fathers as having many more favorable characteristics than most men. Daughters of divorcées viewed their fathers, husbands, and most men as having predominantly undesirable characteristics. Their attitude could be characterized as "Men are no damned good; they never have been and they'll never change." In contrast, girls from nuclear families viewed men in general and their husbands and fathers as good but not perfect, as having a substantial number of favorable attributes but also some flaws.

The appraisals of the husbands on the basis of interviews, tests, and direct observations to some extent confirm the wives' opinions. The husbands of the daughters of divorcées were less educated, had less stable employment records, and were more frequently involved in problems with the law than were the other two groups of husbands. In addition, they felt more ambivalent or hostile toward their wives and infants and were less emotionally mature and more impulsive and self-centered. In contrast, daughters of widows tended to marry husbands with more education or a higher vocational status than their parents had. These men were self-controlled almost to the point of being too inhibited. They were also nurturant, ambitious, concerned about social approval, and conventional, and they maintained stereotyped views of male and female roles. One interviewer characterized them as "repulsively straight." The results of this study indicated that the effects of fathers' absence on their daughters' interactions with males were long-lasting and extended even into their marital choices. These girls seemed to select mates who were similar to their images of their fathers, whereas girls from nuclear families were less constrained by their relationships with their fathers in their choice of husbands. These couples were studied in the first year following marriage. It might be questioned how long the marriages can be sustained for either of these groups of girls from mother-headed families.

It should be noted that the girls from mother-headed families in this study came from rather extreme samples in which not only was there no father but there were no other males in the house, including brothers, and in which the mother had not remarried. These girls could be considered deprived of contact with males in general, rather than just deprived of a father. Such marked differences might not be obtained with less severe conditions of male absence.

In reviewing the findings from these studies on sex-role typing and heterosexual relations, it is again apparent that a mother's functioning and her responses to her situation and her child play a focal role in mediating the adaptation of children in mother-headed families.

The Development of Self-Control

The development of self-control is the second area of personality and social development in which there has been extensive study of children in single-parent families.

Boys in mother-headed families have been found to be more antisocial and impulsive, less self-controlled, less able to delay immediate gratification, and more rebellious against adult authority figures than are boys in nuclear families (Douvan and Adelson, 1966; Hetherington et al., 1977; Mischel, 1961; Santrock, 1970). These characteristics are manifested in a wide range of populations and situations.

In an extensive study of seventh-grade white children (Hoffman, 1970), boys from mother-headed families in contrast to boys from nuclear families had less-well-internalized standards of moral judgment. They tended to evaluate the seriousness of an act according to the probability of detection or punishment rather than in terms of interpersonal relations and social responsibility. They were rated by teachers as more aggressive and less willing to conform to rules or to show consideration for others. Following transgressions, boys from mother-headed homes showed little guilt and were unwilling to accept blame for their own behavior. Instead of accepting responsibility or trying to rectify the situation, these children responded in an immature fashion, denying that they performed the act, crying, making excuses, or blaming others. In this study, no differences were found between girls from single-parent or nuclear families.

Similar results have been found with older subjects. Siegman (1966), in a study of first-year law and medical students, found that males who were without a father for at least one year from age 1 through 4 scored higher on self-reported antisocial behaviors, such as parental disobedience, property damage, and drinking, than did boys from nuclear families. Suedfeld (1967) found that Peace Corps volunteers who were without a father for at least five years before their 15th birthday tended to be among those volunteers who returned prematurely because of adjustment or conduct problems.

In addition, although research on juvenile delinquency is plagued by methodological difficulties, most reviewers conclude that children from mother-headed families are overrepresented in delinquent groups (Biller, 1976; Finer, 1974; Lynn, 1974). Herzog and Sudia (1970) have rallied an impressive amount of evidence that this relationship is more strongly related

> not primarily to father's absence but rather to stress and conflict within the home, inability of the mother to exercise adequate supervision, depressed income and living conditions (including exposure to

unfavorable, neighborhood influences), the mother's psychological and behavioral reaction to separation from her spouse as well as to the social and economic difficulties of her situation as a sole parent, and community attitudes toward the boy and family. (p. 154)

Further evidence that the effects are not attributable simply to unavailability of a father is found in the previously cited study of divorce, in which fathers maintained visitation rights and had some contact with their preschool children (Hetherington *et al.*, 1977). When there was agreement in child-rearing, a positive attitude toward the spouse, and low conflict between parents and when the father was emotionally stable, frequent visitation by the father was associated with more positive adjustment and self-control in the child. When there was disagreement and inconsistency in attitudes toward the child or when the father was poorly adjusted, frequent contact with the child was associated with poor mother-child functioning and disruptions in the children's behavior. Again we see that it is not a simple matter of father availability and amount of contact, but it is the quality of the contact that counts.

These factors are undoubtedly all important in the lack of self-control and high incidence of antisocial behavior that is often found in children in single-parent families, however, they do not adequately account for two notable findings in this literature. The first is that differences in self-control and acting-out behavior between children in nuclear and single-parent families are found for boys but are less marked or are not obtained for girls (Gurin, Veroff, and Feld, 1960; Langner and Michael, 1963; Hetherington *et al.*, 1977; Hoffman, 1970; Nye, 1957; Rosenberg, 1965; Santrock, 1975). The second is that when differences occur between children in single-parent and nuclear homes, they are most likely to be found in children whose parents have been divorced rather than in those in which the father has died.

Studies of both clinic and nonclinic samples of children have shown that children from divorced parents are more likely to manifest their disturbance in acting-out, aggressive, antisocial behavior, whereas children whose fathers have died are more likely to show depression, neurotic symptoms, anxiety, or habit disturbances (Felner, Stolberg, and Cowan, 1975; Hetherington, 1972;

Santrock, 1975; Tuckman and Regan, 1966). In addition, girls are less likely to respond in an antisocial aggressive manner to marital disruption or to their parents' divorce than are boys (Hetherington *et al.*, 1977; Rutter, 1971; Tuckman and Regan, 1966). Both male and female children from homes in which parents have divorced or the father has died would be exposed to many of the stressful familial and cultural factors described by Herzog and Sudia. What explanation can be offered for the differences in outcome of divorce and death of a father? It has been suggested that the child adopts the predominant behavior exhibited by his parents during a crisis and that this becomes a guiding framework for his later behavior and coping. In the case of divorce—particularly in one that has been preceded by considerable acrimony and conflict—anger, aggression, and hostility are the behaviors displayed by the parents and adopted by the child. In the case of death, it is the parent's mourning, depression, and withdrawal that is introjected (Felner *et al.*, 1975).

In addition to the acrimony and conflict likely to accompany divorce but not the death of a father, the greater social stigma associated with divorce, anger at being abandoned by the father, and the more negative image of the father likely to be communicated by the mother may all contribute to greater stress and resentment in the divorced family members. Divorcées are also likely to be younger than widows and to have younger children. Young children have been shown to create more stress in mothers than do older children. Finally, widows seem to have more extensive support systems, particularly from the spouse's family, than are available to divorcées (Hetherington, 1972). These factors may all contribute to greater duress and frustration in divorced families and to acting-out behavior in children.

Why do boys respond to divorce with more antisocial, aggressive, uncontrolled behavior than do girls? It may be that the greater aggressiveness frequently observed in boys and the greater assertiveness in the culturally proscribed male role necessitates the use of firmer, more consistent discipline practices in the control of boys than of girls. Boys in nuclear families are less compliant than girls, and children are less compliant to mothers than to fathers. It could be argued that it is more essential for boys that they have a male model to imitate who exhibits self-controlled ethical behavior, or that the image of greater power and authority vested in the father is more critical in the control of boys, who are socialized to be more aggressive. Is this the main factor in the greater aggressive antisocial behavior in boys in divorced families? Let us go back once again to the longitudinal study by Hetherington *et al.* (1977) of the two years following divorce. As we noted earlier, the parenting practices and control of the divorced mothers deteriorated over the course of the first year following divorce, although they dramatically improved during the second year. Poor parenting was most apparent when divorced parents, particularly divorced mothers, were interacting with their sons. They exhibited less positive behavior, affection, and affiliation with their sons than with their daughters. In addition, they gave more threatening commands that they didn't systematically enforce with sons.

Divorced mothers may have given their children a hard time, but mothers, especially divorced mothers, got rough treatment from their children. As was previously remarked, children are more likely to exhibit oppositional behavior to mothers and compliance to fathers. Children more frequently made noncompliant, negative, complaining demands to the mother than to the father. Boys were more oppositional and aggressive; girls were more whining, complaining, and compliant. Divorced children showed an increase in dependency over time and exhibited less sustained play than children in intact families. The divorced mother was harassed by her children, particularly her sons. In comparison with fathers and mothers in intact families, her children in the first year didn't obey, affiliate, or attend to her. They nagged and whined, made more dependency demands, and were more likely to ignore her. The aggression of boys with divorced mothers peaked at one year following divorce, then dropped significantly but was still higher than that of boys in nuclear families two years after divorce. Divorced mothers described their relationship with their child one year after divorce as "declared war," a "struggle for survival," "the old Chinese water torture," or "like getting bitten to death by ducks." As we found in the divorced parents' behavior, one year following divorce seemed to be the period of maximum negative behaviors for children, and great improvement occurred by two years, although the negative behaviors were more sustained in boys than in girls. The second year appears to be a period of marked recovery and constructive adaptation for divorced mothers and children.

Poor parenting was associated with aggression, impulsivity, and lack of control in boys both in the home and in a preschool setting. The importance of the maternal parenting practices relative to paternal practices increased markedly over the two-year period.

Who is doing what to whom? Mothers and sons seemed to be involved in a coercive cycle. The lack of management skills of the mother accelerated the son's aversive behavior, for which she was the main instigator and target. This led to increased feelings of helplessness, anxiety, depression, and incompetence in the mother, which was associated with increased coercion in her parenting skills, and the cycle continued. How can we intervene in this cycle? Although dynamic insight-oriented psychotherapy was not found to be an effective form of intervention, therapy focused specifically on parenting was. In a subgroup of mothers who were involved in programs focused on the improvement of parenting skills and in which 24-hour telephone contact with the parent trainer was available, both the divorced mother and the child demonstrated improved adjustment. This was manifested in an improvement in self-esteem, happiness, and feelings of competence in the mothers and in increased self-control in the children. The coercive cycle was broken.

Another longitudinal investigation that studied families immediately after divorce and 12 to 18 months later found different reactions of children to divorce as a function of age (Wallerstein and Kelly, 1975). In general, in accord with the previous findings, family relations deteriorated over the first year following divorce. The quality of the mother-child relationship and the emotional adjustment of the mother seemed more important in contributing to the behavior problems of younger than of older children. In fact, many of the school-aged children actually seemed to be providing more emotional support for their mothers than they were receiving. When divorced mothers were so preoccupied with their own intense emotional problems and anger that they were unable to help or comfort their children, or when there was a breakdown in their parenting capacity, their children were particularly vulnerable to emotional disturbance and acting-out behavior.

CONCLUSION

Children living in mother-headed single-parent homes appear to be at higher risk for disruptions in cognitive, emotional, and social development than are children in nuclear families. These developmental disruptions do not seem to be attributable mainly to father absence but to stresses and a lack of support systems that results in changed family functioning for the single mother and her children. Although the stresses and demands of single-parent life are more incessant and burdensome than those in nuclear families, many mothers are able to deal with them in a constructive and rewarding manner. When maternal responses are positive and adaptive, the children are likely to demonstrate intellectual and social competence.

An increasing number of children are going to grow up in single-parent mother-headed families. It is critical to develop social policies and intervention procedures that will reduce stresses and develop new support systems for single-parent families in order to offer these families more constructive and fulfilling life-styles.

REFERENCES

Anspach, D. F. Kinship and divorce. *Journal of Marriage and the Family,* 1976, *38,* 330-343.

Apperson, L. B., and McAdoo, W. G., Jr. Parental factors in the childhood of homosexuals. *Journal of Abnormal Psychology,* 1968, *73,* 201-206.

Bach, G. R. Father-fantasies and father-typing in father-separated families. *Child Development,* 1946, *17,* 63-80.

Bane, M. J. Discussion paper, health, education, and welfare. Policy toward children, youth, and families, 1977

Barclay, A. G., and Cusumano, D. Father-absence, cross-sex identity and field-dependent behavior in male adolescents. *Child Development,* 1967, *38,* 243-250.

Bene, E. On the genesis of female homosexuality. *British Journal of Psychiatry,* 1965, *3,* 815-821.

Bieber, I. *Homosexuality: A Psychoanalytic Study.* New York: Basic Books, 1962.

Biller, H. B. *Paternal Deprivation: Family, School, Sexuality, and Society.* Lexington, Mass.: D. C. Heath, 1974.

Biller, H. B. The father and personality development: Paternal deprivation and sex-role development. In M. D. Lamb (Ed.), *The Role of the Father in Child Development*. New York: Wiley, 1976, pp. 89-156.

Biller, H. B., and Bahm, R. M. Father-absence, perceived maternal behavior, and masculinity of self-concept among junior high school boys. *Developmental Psychology*, 1971, 4, 178-181.

Biller, H. B., and Borstelmann, L. J. Masculine development. An integrative review. *Merrill-Palmer Quarterly*, 1967, 13, 253-294.

Blanchard, R. W., and Biller, H. B. Father availability and academic performance among 3rd grade boys. *Developmental Psychology*, 1971, 4, 301-305.

Carlsmith, L. Effect of early father absence on scholastic aptitude. *Harvard Educational Review*, 1964, 34, 3-21.

Cox, M. The Effects of Father Absence and Maternal Employment on the Development of Children. Unpublished doctoral dissertation, University of Virginia, 1974.

Crescimbeni, J. Broken homes do affect academic achievement. *Child and Family*, 1965, 4, 24-28.

Deutsch, M., and Brown, B. Social influences in Negro-white intelligence differences. *Journal of Social Issues*, 1964, 20, 24-35.

Douvan, E., and Adelson, J. *The Adolescent Experience*. New York: Wiley, 1966.

Evans, R. B. Childhood parental relationship of homosexual men. *Journal of Consulting and Clinical Psychology*, 1969, 33, 129-135.

Felner, R. D., Stolberg, A., and Cowan, E. L. Crisis events and school mental health referral patterns of young children. *Journal of Consulting and Clinical Psychology*, 1975, 43, 305-310.

Finer, M. *Report of the Committee on One-Parent Families*. London: Her Majesty's Stationery Office, 1974, 2 vols. (Can be ordered from: The Government Bookshop, H.M.S.O., P.O. Box 569, London SE19NH, England).

Fisher, S. F. *The Female Organism: Psychology, Physiology, Fantasy*. New York: Basic Books, 1973.

Funkenstein, D. H. Mathematics, quantitative aptitudes and the masculine role. *Diseases of the Nervous System*, 1963, 24, 140-146.

Glasser, P., and Navarre, E. Structural problems of the one-parent family. *Journal of Social Issues*, 1965 (January), 98-109.

Gurin, G., Veroff, J., and Feld, S. *Americans View Their Mental Health*. New York: Basic Books, 1960.

Herzog, E., and Sudia, C. E. Children in fatherless families. In B. M. Caldwell and H. Riccuiti (Eds.), Review of Child Development Research, pp. 141-232. Chicago: University of Chicago Press, 1973.

Hess, R. D., Shipman, V. C., Brophy, J. E., and Bear, R. M. The Cognitive Environments of Urban Preschool Children. Graduate School of Education, University of Chicago, 1968. (ERIC Document Reproduction Service No. ED 039 264.)

Hess, R. D., Shipman, V. C., Brophy, J. E., Bear, R. M., and Adelberger, A. B., The Cognitive Environments of Urban Preschool Children: Follow-up Phase. Graduate School of Education, University of Chicago, 1969. (ERIC Document Reproduction Service No. ED 039 270.)

Hetherington, E. M. The effects of familial variables on sex-typing, on parent-child similarity and on imitation in children. *Minnesota Symposium on Child Development*, 1967.

Hetherington, E. M. Effects of paternal absence on personality development in adolescent daughters. *Developmental Psychology*, 1972, 7, 313-326.

Hetherington, E. M., Cox, M., and Cox, R. The aftermath of divorce. In J. H. Stevens, Jr., and Marilyn Matthews (Eds.)., *Mother-Child, Father-Child Relations*. Washington, D.C.: National Association for the Education of Young Children, 1977.

Hetherington, E. M., Cox, M., & Cox, R. The Impact of Divorce on Children: A Longitudinal Study. Unpublished paper, 1978.

Hoffman, M. L. Moral development. In P. Mussen (Ed.), *Handbook of Child Psychology*. New York: Wiley, 1970.

Jacobson, G., and Ryder, R. G. Parental loss and some characteristics of the early marriage relationship. *American Journal of Orthopsychiatry*, 1969, 39, 779-787.

Johnson, M. M. Fathers, mothers and sex typing. *Sociological Inquiry*, 1975, 45 (1), 15-26.

Lamb, Michael The effects of divorce of children's personality development. *Journal of Divorce*, 1977, 1.

Langner, G. S., and Michael, S. T. *Life Stresses and Mental Health*. New York: Free Press, 1963.

Lessing, E. E., Zagorin, S. W., and Nelson, D. WISC subtest and IQ score correlates of father absence. *Journal of Genetic Psychology*, 1970, 117, 181-195.

Lynn, D. B. *The Father: His Role in Child Development*. Belmont, Calif.: Brooks/Cole, 1974.

Lynn, D. B., and Sawrey, W. L. The effects of father-absence on Norwegian boys and girls. *Journal of Abnormal and Social Psychology*, 1959, 59, 258-262.

Marsden, D. *Mothers Alone: Poverty and the Fatherless Family*. London: Allen Lane, Penguin Press, 1969.

McCord, J., McCord, W., and Thurber, E. Some effects of paternal absence on male children. *Journal of Abnormal and Social Psychology*, 1962, 64, 361 369.

Miller, W. B. Lower-class culture as a generating milieu of gang delinquency. *Journal of Social Issues*, 1958, 14, 5-19.

Mischel, W. Father-absence and delay of gratification. *Journal of Abnormal and Social Psychology*, 1961, 62, 116-124.

Mueller, C. W., and Pope, H. Marital instability: A study of its transmission between generations. *Journal of Marriage and the Family*, 1977, 39 (1, February), 83-94.

Mussen, P. H., and Rutherford, E. E. Parent-child relationships and parental personality in relation to young children's sex-role preferences. *Child Development*, 1963, 34, 589-607.

Nelson, E. A., and Maccoby, E. E. The relationship between social development and differential abilities on the Scholastic Aptitude Test. *Merrill-Palmer Quarterly*, 1966, 12, 269-289.

Nye, F. I. Child adjustment in broken and in unhappy unbroken homes. *Marriage and Family Living*, 1957, 19, 356-360.

Oshman, H. P. Some Effects of Father Absence upon the Psychological Development of Male and Female Late Adolescents: Theoretical and Empirical Considerations. (Doctoral Dissertation, University of Texas at Austin, 1975.) Dissertation Abstracts International. 36, 919B-920B. (University Microfilms No. 75-16, 719).

Pearlin, L. I., and Johnson, J. S. Marital Status, Life-Strains and Depression. Unpublished manuscript, 1975.

Pope, H., and Mueller, C. W. The intergenerational transmission of marital instability: Comparisons by race and sex. *Journal of Social Issues*, 1976, *321, (Winter)*, 149-166.

Rees, A. H., and Palmer, F. H. Factors related to change in mental test performance. *Developmental Psychology Monograph*, 1970, *3*, 1-57.

Rosenberg, M. *Society and the Adolescent Self-Image*. Princeton, N.J.: Princeton University Press, 1965.

Rutter, M. Parent-child separation: Psychological effects on the children. *Journal of Child Psychology and Psychiatry*, 1971, *12*, 233-260.

Saghir, M. T., and Robbins, F. *Male and Female Homosexuality*. Baltimore: Williams & Wilkins, 1973.

Santrock, J. W. Paternal absence, sex-typing, and identification. *Developmental Psychology*, 1970, *2*, 264-272.

Santrock, J. W. Relation of type and onset of father-absence and cognitive development. *Child Development*, 1972, *43*, 455-469.

Santrock, J. W. Father absence, perceived maternal behavior, and moral development in boys. *Child Development*, 1975, *46*, 753-757.

Santrock, J. W., and Wohlford, P. Effects of father absence: Influences of, reason of, and onset of absence. *Proceedings of the 78th Annual Convention of the American Psychological Association*, 1970, *5*, 265-266.

Sears, P. S. Doll-play agression in normal young children: Influence of sex, age, sibling status, father's absence. *Psychological Monographs*, 1951, *65* (Whole No. 6).

Sears, R. R., Rau, L., and Alpert, R. *Identification and Child Rearing*. Stanford, Calif.: Stanford University Press, 1965.

Shinn, M. Father absence and children's cognitive development. *Psychology Bulletin*, 1977, *85*, 295-324.

Siegman, A. W. Father-absence during childhood and antisocial behavior. *Journal of Abnormal Psychology*, 1966, *71*, 71-74.

Solomon, D., Hirsch, J. G., Scheinfeld, D. R., and Jackson, J. C. Family characteristics and elementary school achievement in an urban ghetto. *Journal of Consulting and Clinical Psychology*, 1972, *39*, 462-466.

Spicer, J. W., and Hampe, G. D. Kinship interaction after divorce. *Journal of Marriage and the Family*, 1975, *37*, 113-119.

Stack, Carol B. Black kindreds: Parenthood and personal kindreds among urban blacks. *Journal of Comparative Family Studies*, 1972, 194-206.

Suedfeld, P. Paternal absence and overseas success of Peace Corps volunteers. *Journal of Consulting Psychology*, 1967, *31*, 424-425.

Sutton-Smith, B., Rosenberg, B. G., and Landy, F. Father-absence effects in families of different sibling compositions. *Child Development*, 1968, *39*, 1213-1221.

Thompson, N. L., Schwartz, D. M., McCandless, B. R., and Edwards, D. A. Parent-child relationships and sexual identity in male and female homosexuals and heterosexuals. *Journal of Consulting and Clinical Psychology*, 1973, *41*, 120-127.

Tuckman, J., and Regan, R. A. Intactness of the home and behavioral problems in children. *Journal of Child Psychology and Psychiatry*, 1966, *7*, 225-233.

U.S. Bureau of the Census. Household and Family Characteristics. Current Population Reports Series P-20, No. 291, March. Washington, D.C.: U.S. Government Printing Office, 1976.

Wallerstein, J., and Kelly, J. The effects of parental divorce: Experiences of the preschool child. *Journal of the American Academy of Child Psychiatry,* 1975, *14,* 600-616.

Weiss, R. S. A preliminary examination of potential contributions of quality of life data to an understanding of single parenting, 1977, unpublished manuscript.

Wohlford, P., Santrock, J. W., Berger, W. E., and Liberman, D. Older brothers' influence on sex-typed, aggressive, and dependent behavior in father-absent children. *Developmental Psychology,* 1971, *4,* 124-134.

Zajonc, R. B. Family configuration and intelligence. *Science,* 1976, *192,* 227-236.

4

Practical Considerations

Ira D. Glick

Dr. Hetherington's work has provided us with data on a new family form—mother-led families—in a way that previously has not been available to most professionals working with individuals or families.

Let me offer several caveats about her research. First, there is an enormous difference between what people report and what actually happens. The kind of data gathering in her chapter—depending in large measure on verbal reports—has significant limitations. Second, we have to be careful of the rubrics *mother-headed* or *father-headed* families. Such rubrics are a way of characterizing things, but they tend to throw you off in terms of what actually is occurring in the family system. If there is any one point I want to emphasize, that is it. You've got to look at what's in the system—not just what the mother is doing in the household, or just a father popping in and out of the house. Quality of data is important, but a variety of other kindred influences on the family

Ira D. Glick • Cornell University and New York Hospital Medical Center; Payne Whitney Clinic, New York, New York.

also determine what happens to the kids and contributes to successful or unsuccessful coping after a divorce.

In terms of clinical application, I would start with a careful appraisal of the actual family situation in the single-parent mother-headed household. Systematically, I would look at that and recall the successful coping strategies and problem areas that Dr. Hetherington discussed. Then I would try to determine what deficiencies I could see and what kind of family system changes are needed to make the situation work best.

There is a bias that the single parent is going to do terribly and that the kids will too. We *don't* know that yet with any certainty. I don't want to sound like a total cynic, saying that we don't know anything, but we don't know a lot about child development and outcome yet for either intact or single-parent families.

After making a careful diagnosis of the deficiencies and problems in the family system, what can one do about it? One of the results of the family therapy movement has been an enormous increase in family therapy both for the single parent and for the divorcing family. Family therapy is a useful intervention that, many believe, offers significant advantages over individual or group therapy.

After evaluating the family system, the therapist formulates goals and starts focusing on the system. You have to make some hard choices when you're doing family system intervention rather than individual intervention. Individual intervention is often based on what's best for that particular mother or father. Family system intervention is based on what's best for the family system. Now, having emphasized in my discussion of Dr. Kohn's chapter not to impose your values on the family, let me say that a family therapist *has to* make some value judgments based on clinical experience and what is in the literature. The therapist must try to elucidate the options for a particular family. The family then makes the choices.

Clearly, the data from Dr. Hetherington's paper suggest that one of the primary goals is to help *a mother* cope in a constructive and fulfilling way—for example, helping her to reduce the guilt she might feel about divorce, helping her find personal satisfaction in a social life, perhaps helping her to find work to help alleviate economic problems. Second, Dr. Hetherington's data strongly imply that a family system therapist would try to involve *the father* in this one-household family, provided much of the discord between the parents has been resolved (if it hasn't, more disruption can occur). As Dr. Hetherington points out, the benefits can be enormous in terms of the father's role in successfully helping the children to cope. Third, the family therapist wants to work with the *children*. You must help the children cope with the stress of the parents' moving apart. Often the kids blame the parent initiating the divorce. What has got to be made clear to the children—and it's often not clear to the therapist—is that people mostly try to do the best they can in difficult situations and that often divorce can be a better solution than staying in a difficult marriage. Somehow, this point must be communicated to the children and communicated on a level they can understand, regardless of age. A practical change that a therapist can suggest is to help the children take on some responsibility for family work, for example, washing dishes, and setting the table.

A case example might make this more clear. In the L family, the mother and father divorced after a 10-year marriage. There was a great deal of bitterness about the divorce, and the mother didn't want the father to do much caring for the children. The father was angry with the mother for instituting the divorce. Because of this, he saw his two young children only on rare occasions. However, he was interested in them. The mother was going out very little, staying home with frequent headaches. The task of the family therapist was to help the mother socialize more and to help her find a job. The children were asked to help in setting the table and cleaning up after meals, taking care of their rooms, and so on. Family meetings were held with both parents. The father was encouraged to take the children five times as much as he previously had. The result was that the mother's headaches diappeared and she found a job and began dating. The father was less angry with the children and found himself happier. Both children, who had been having trouble in school, improved their performance there.

Dr. Hetherington's paper has vividly made the point that if you have to *work* at making an intact marriage successful, you have to *work twice as hard* to make a single-parent household function successfully.

Practical Considerations

Carol F. Hoover

Dr. Hetherington's presentation was certainly very stimulating. It brought up almost immediately a number of possible clinical applications, one of which was a comforting therapeutic use of the report that a year or two after a divorce, some of the acute problems in child-rearing tend to level off. I know people always used to say that the first year of a marriage was the hardest; I guess now we can say that the first year of a divorce is the hardest.

Also, the development of a support network by divorced or widowed women might be encouraged, with an earlier emphasis. It is important to help lone women with children to establish relationships quickly with other individuals whom they can depend on or with whom they can be aligned.

Oftentimes, research doesn't specify whether mother-headed families include other people besides the mother and the children. Dr. Hetherington, in one of her studies, did indicate that there were no men present in the household, not even any male children. In general, however, figures of this kind are very hard to get.

Carol F. Hoover • National Institute of Mental Health, Bethesda, Maryland.

Even with Dr. Sussman's excellent figures, you would have had to add here and subtract there and combine this figure with that to arrive at much of an estimate. I would guess that many women who head single-parent families also have what you would call the *expanded family*—that is, related or sometimes nonrelated individuals who live in the same household.

If I recall correctly a study conducted of black families in the New York metropolitan area, behavior disorders were significantly more present in boys who lived in families where only one adult was present (the mother), but it did not much matter whether the second adult was the child's father, another male, or another female. A grandmother or an aunt or an uncle or a friend served just about as well (statistically speaking), when added to the mother-headed family, as the child's own father.

Another study by Dr. Eiduson's group in California compared the child-rearing practices of unmarried couples, commune-living couples, conventionally married pairs, and single women with illegitimate children.[1] Some of the single mothers later lived with the child's father, or with their own kin, or with unrelated persons, as well as alone. Moreover, half of the unmarried mothers had planned their pregnancies, and one-third of them had deliberately intended to bring up the child without a father. That was the way they wanted to raise their children.

This raises an important methodological question about research with divorced parents. How many mothers (or fathers) chose to dispense with the domestic interference of a second parent? How many prefer single blessedness to marriage? By contrast, who of them feel helpless and abandoned without husband or wife to help with child care? We need to examine how the preference for singlehood, or lack of it, may influence the development of problems, or the absence of them, in children of these separations.

I was very much interested in the report by Dr. Hetherington concerning the daughters of widows, compared to the daughters of divorcées, who went to a recreation center dance. As Dr. Hetherington said, this report can be interpreted in a number of

---

[1]For more details on this work, see Chapter 5.

different ways. One thought I had was that to some extent, the girls may have been copying their mothers. That is, the divorcées perhaps had more sense of having chosen their own status, or at least having participated in a choice, and were more actively initiating new contacts. The children of widows may, like their mothers, have felt more helpless; something had happened to them over which their mothers had no control.

When mother-headed families are compared to intact families, the statistical compilation of the former may represent an average of vastly different situations: unhappy, embittered divorcées; women who manage with fair competence but not too happily without husbands; and women who much prefer their present existence to the former married state.

I think we need more current data on such points. It would certainly be difficult to establish who initiates divorces; indeed, it is my observation that most divorces in the end are rather mutual, although one partner may be more enthusiastic than the other. However, it should be rather simple and routine, preliminary to other research questions, to ask formerly married people whether they feel happier unmarried than married, or the reverse, or just about the same; and whether, all other things being equal, they would prefer remarriage. We need to know if women today feel differently than men about living apart. Does having children make a more profound difference to divorcing partners of one sex than to the other? Perhaps the experience of single parenthood is quite different at varying social and economic levels. I suspect that there is still so much value placed on marriage in our culture—and women continue to appear so dependent on their husbands—that it is virtually taken for granted that formerly married women must be less than happy. It seems almost an affront that a woman should prefer to be a carefree divorcée rather than a patient Griselda, still more of a shock that she might enjoy being a merry widow more than a faithful wife. With children yet! The woman who does not bed down legally with one man may be regarded even today as setting a poor example for her offspring, no matter how stable and enriching her love life. The stereotypes of wifehood versus promiscuity versus lonely abandonment die hard.

It is also assumed, usually, that two parents must be better than one, unless extreme marital conflict exists in an intact marriage. Perhaps so, but the unmarried women who choose to get pregnant and raise children without a husband and the single women who adopt children indicate that not everyone agrees with this thesis any longer.

As a therapist who has dealt with uncounted couples and families, I have been struck by how many of them report that most of their marital arguments are about the children. Well, one of the delights of raising a child alone is that there are no arguments with the other parent. No battles about discipline. What *you* say goes! If there are visits from the other parent, there may be rivalries for affection, but usually after a year or so, the noncustodial parent stops seeing the children so often. Under these circumstances, the children can see quite clearly which parent is most involved and adjust their attachments accordingly. Fantasies of the rescuing other parent may of course persist, but in general, reality triumphs.

We need more comparative studies of single-fathered and single-mothered families. At present, fathers who act as single parents represent rather unusual cases. Some of them are vindictive betrayed husbands who are determined to use their children for revenge; others are men whose identity as fathers assumes more importance to them than for most men in our current society; some can't bear to lose a close relationship with their children while knowing that their wives have a superficial or cool relation with the children at best; and still others have become used to looking after the children because of their wives' incapacity and find that they have no more trouble doing so than the average woman who acts as single parent. I do a good deal of work with men whose wives have been recurrently hospitalized for psychiatric illness over many years. Some of those men have gradually become so expert at managing a household and looking after the children that they are just as good as the best women in managing the single-parent status. They get the children off to nursery school in the morning, and then they go to work, come home, fix supper, and get the children ready for bed. It's all done with a great deal of aplomb.

I think that what people are now calling the Pierre Trudeau phenomenon will probably increase in the next few years, especially as the courts lose their peculiar and rather late-instituted preference for mothers as custodians (it is primarily since 1900 that this preference has come about). We should see quite a lot of ordinary husbands and wives agreeing on who is to retain custody of the children without there being some kind of guilt implied if the wife isn't the one who takes them. A man may choose to raise his children rather than pay child support—especially if he is the one getting the most money. The wife may prefer complete independence as she climbs the bottom rungs of a career ladder or gets herself newly-educated.

If we then compare male-parented and female-parented families, we may compile more data on how much disturbance in children, for example, is attributable to the disruptions of divorce or the loss of a parent, how much is due to the absence of a second adult in the home, how much is due to lack of sex-role models of both varieties, and so on.

We can also continue to take a look at how children raised with different child-rearing philosophies are different. Although the California study referred to earlier had originally intended to compare communes with three other forms of child-raising situations, researchers found that the wide variation in each of the settings was confounding the whole study. There were tremendous variations in communes, within so-called traditional marriages, in looser partnerships, and among unmarried mothers. Some communes, for example, stressed traditional sex-role models and family togetherness; certain ones had special sect or cult religious values; others left children in a fluid group arrangement, gave each of the little ones a bottle, and let them choose to cuddle with their peers. Eventually, the California researchers decided that the cohort comparisons were not nearly so important as determining how each particular parent or couple of parents had decided to raise their child.

In therapeutic work with families, it often appears that the most pathological situations are those where no one can separate from anyone else except under the most damaging circumstances: where children are stuck with their parents, and the parents with each other, and a dominant grandparent often thrown in for good measure, from whom the parents have found it impossible to separate emotionally. Even a therapist sometimes finds it hard to visualize how any one of those vulnerable people could manage by herself or himself. Some of these groups may be openly conflict-ridden, others conflict-repressive.

What makes the difference between these poisonously glued families and those where a joyful mutual dependence exists with a fair measure of autonomy, and where separation, if need be, can represent a new and sometimes painful but also productive phase in their several lives? That is a question that clinical investigators must try to answer.

5

Emergent Families of the 1970s: Values, Practices, and Impact on Children

Bernice T. Eiduson

PLURALISM IN THE CONTEMPORARY AMERICAN FAMILY

The less-than-intact family is not new to the clinician. We have all learned and had confirmed through years of clinical experience that societal disorganization and individual pathology often go hand-in-hand with family disorganization and family pathology. We have had the "aha experience" when searching the record of a juvenile delinquent to find that he essentially was homeless, uprooted from biological parents at an early age; we have shaken our heads knowingly when the father deserts and a child becomes apathetic and withdrawn; we have expected, with a history of early separation between mother and child, a drifting, rootless, unattached youngster. We have had compelling evidence through our

---

Bernice T. Eiduson • University of California, Los Angeles, California.

experience with individuals who develop deviantly that indeed, the family is the primary socialization agent for the child. Experience has sensitized us, perhaps too much so, to what the stable, rooted, intact nuclear family ideally can and should be providing, not only to the child but to every member. Therefore, when modification in the traditional nuclear family unit has taken place in the past, it is not surprising that clinicians expect the worst and tool up for the resulting personal and societal traumas.

However, our view of the nuclear family has been shaken during the late 1960s and early 1970s, when young people who had grown up in traditional two-parent nuclear families, who were Caucasian and affluent, and many of whose parents were educated professionals and executives, vociferously attacked the inadequacies of their traditional nuclear families (Eiduson, Cohen, and Alexander, 1973). They cited their parents' unhappiness, psychosomatic problems, alcoholism, and self-doubts within the framework of a good and stable home (Keniston, 1965). The values that they espoused as young adults seemed unable to be embraced in the traditional two-parent nuclear unit in which they had grown up (Roszak, 1969; Cooper, 1970; Otto, 1970). They stated that absent from the traditional two-parent nuclear family were such values (Eiduson et al., 1973; Weisner et al., 1976), as:

1. Humanism; with its opportunities for closeness, intimacy, and significant emotional relationships.
2. A natural organic perspective that desires to see man as at one with the environment.
3. A move away from dependence on technology and invention, and the intellectual perspective that uses only rational and objective sources of knowledge, toward an interest in including diverse and personal sources of knowledge, such as sensory experience, intuition, and astrology, as alternate ways of problem solving.
4. A self-help philosophy where the individual has more control of his own fate and is less dependent on established institutions.
5. A withdrawal from materialism, possessiveness, and acquisitiveness as important values in life, in favor of self-fulfillment, creative expression, and respect for individual differences.

6. An encouraging of roles within the family without regard for stereotyped conceptions of sex or age.
7. More immediate gratification, setting of day-to-day goals, and appreciation of the here-and-now.

Some of the family forms that they saw as holding out to them the potential for embodying such values were living groups, social-contract rather than legally contracted marriages, communities, religious "families," triads and polyandrous arrangements, and families with single mothers as head of household.

These family forms were not without previous history in the United States (Berger, 1967; Shorter, 1975). The commune, for example, is not in itself a new experience in this country (Gollin, 1967; Zablocki, 1971; Kanter, 1972). Some Americans have always sought a new start as an expression of dissent against the status quo. Causes of their dissent and the ways they chose to organize their new communal existence varied in past generations as it does for today's communards. Some were based upon religious convictions and others upon economic idealism and a rejection of gross economic inequalities. Some rebelled against authority and attempted to establish a model of governing based upon an absence of central authority; others sought a clear structuring along strict lines of hierarchical authority and ultimate rejection from the community of those who would not yield to such authority. However, these family variants, and others in the past, had almost exclusively been adopted by default, so that they were not seen merely as different but also as deviant (Wright, 1965; Deutsch, 1967; Williams, 1972; Furstenberg, 1976).

In the 1960s, however, young people adopted family variants by choice; they rejected what had been the modal and preferred family style and forced us, as professionals, to reevaluate how one should define a family (Tannenbaum, 1969; Skolnick and Skolnick, 1971; Bengston and Laufer, 1974). Could a family be comprised of both kin and nonkin members? How then would they function and relate to each other? What role do goals or ideologies play in influencing how a family operates? Can families successfully recast the traditional roles of adults and so restructure the experiences, responsibilities, and opportunities for every family member?

It appeared by the 1970s that social-contract, single-parent, and some of the communal households were viable family entities once the phase of experimentation with diverse units had passed (Rains, 1971; Delora and Delora, 1972; Kanter, 1972). These emergent family forms warranted close scrutiny for numerous reasons:

1. Young people who were attracted to the restructuring of the family were of child-bearing age. Thus, they were likely to have young children whom they would want to rear in accordance with their values. What would this mean in terms of child-rearing practices and in terms of developmental outcomes of the children?

2. Although the original family forms were experimental and appeared to be perhaps a transitory social phenomenon, there was increasing evidence from census data and from unsystematic journalistic reports that some of the alternative forms not only were becoming a visible part of American society but were increasing and were attracting not only young adults but other strata of the population as well (Lipetz and Davis, 1972; Rogers, 1972; Glick, 1975).

3. Even more obvious than the spread of nonconventional family forms was the diffusion into mainstream society of many of the values identified with the alternative philosophy (Eiduson, Cohen, and Alexander, 1973). One has only to look at the mass media, at movies, television, art, or fashion, to become aware of how these aspects of society that are the most responsive to cultural change have resonated to and incorporated some of the values and even the external features of counterculture youth (Miller and Sjoberg, 1973). Therefore, in regard to other aspects of our society that are more slow to respond to changing attitudes, like child rearing or traditional sex-roles, we were anticipating that the same kind of diffusion would ultimately take place.

For these reasons, a careful study of the phenomenon of the changing contemporary young families seemed desirable.

STUDYING EMERGENT FAMILY FORMS

An interdisciplinary group at the University of California at Los Angeles initiated a longitudinal study on 200 families in which the mother was in the third trimester of pregnancy.

We selected 150 children and their families to represent three groups of 50 alternative families (social-contract families, single-mother households, and living groups or communities), and 50 children born into the two-parent traditionally married nuclear family served as a comparison group. The populations were Caucasian parents aged 18-32, primarily of middle-class or stable working-class backgrounds, in accordance with the characteristics of members of the counterculture groups whose ideologies had given rise to contemporary alternative family styles. The mother was pregnant with her first child in two-thirds of the cases, her second in one-third. The aims of the study were to document the child-rearing practices, attitudes, and environmental milieus of alternative families, comparing these with traditional families, and to study the implications for the physical, cognitive, and socioemotional development of the child in the early formative years of life (Eiduson, 1975).

The 150 children and families who comprise the alternative populations were obtained through the indigenous network throughout California. Referrals were sought through Lamaze teachers, physicians who perform home births, ecology centers, and social and educational agencies that alternatives frequent. This "snowball" sample contrasted with the random sample of traditional marrieds, the comparison group, who were obtained through randomly sampling of the list of obstetricians in the AMA directory. Each physician was asked to nominate one woman from his practice who met our criteria. The physician then encouraged her to contact the project.

The research strategy for comparing the lives of children growing up in these four family styles and for evaluating how development proceeds and how development is affected by the differences in environment demands systematic and periodic studies of family and children. Starting with the third trimester, when parents were engaged in the study, an assessment of the socialization processes in the child's family proceeded through interviews, questionnaires, and home visits, permitting naturalistic observations of children, parents and others. In addition, studies of the social and physical milieu of the home provide an ecological focus through a variety of field techniques. To assess the child's development, pediatric and neurological examinations were conducted at birth, home observations initiated at six months, intelligence testing at eight months, and a cognitive, physical, and emotional assessment at one year, etc. This multimethod multitrait research strategy involves administration of standardized and semistandardized instruments and standard experimental situations either in the home setting or in project offices at periodic intervals. Studies will be carried forth until the child completes first grade, so that the impact of family socialization processes on the child's school adjustments can be examined as the child interfaces with mainstream society.

This research approach considers the child the cohort in the study rather than his family. Thus, he is followed, whatever the changes in the family life-style or residence; these latter actually do become study data.

Incentives for participating are of two kinds: first, $5.00 per procedure, to cover the expenses involved; secondly, $80 per year in pediatric services for each child is given any provider of services selected by the parent. In exchange, the project obtains from the service provider codifiable information concerning the nature of the illness, the reason for contact, diagnosis, treatment recommendations, and use of medical services by the family.

THE FAMILY GROUPS

We had hypothesized that the family environments of our four life-style groups would present different functional, intellectual, and relationship settings to the child (Eiduson *et al.*, 1977; Weisner *et al.*, 1976). On the basis of our work, we can offer substantive information on the specific characteristics in which the various family environments differ and are similar, for we have meticulously documented the structure, composition, organization, and neighborhood and community involvement of each household. For example, in one study in which the most conventional and common family unit under study, the traditional marrieds, was compared with the most nonconventional, the living group, the traditional two-parent nuclear unit was found to be a smaller and more restricted social unit, one that was also more stable in terms of residential mobility and of family membership. The traditional family also is maintaining significantly more emotional and physical involvement with own kin. This finding is of interest because it suggests that this variable, relationships with own parents and siblings, which we showed in a previous study (Cohen and Eiduson, 1975) to be one of the few characteristics distinguishing parents who sought an alternative versus a traditional family unit prior to the birth of the child, seems to remain an important difference after the child is born. The traditional family also provides a setting in which the mother herself undertakes more of the total household duties, although her overall workload appears lighter than that borne by her counterpart in the extended group (Weisner and Martin 1976).

Such findings as these permitted us to conceptualize the family structures in terms of some of the within-group differences that had been built into the study by our decision to obtain as much variation in family style in each of our four groups as possible, since there seemed to be no "typical" or representative living group, or single-mother family (Eiduson *et al.*, 1973). Thus, creedal living groups were distinguished from domestic household living groups along ideological as well as certain functional and structural dimensions, such as mobility, stability in caretakers for the project child, and daily routines of the child. Similarly, "circumstantial" social contracts could be differentiated from "committed" social contracts, in terms of the motivations for entering a non-legal family unit; and within the single-mother group, a typology of "nest builders," "post hoc adaptors," and "unwed mothers" appeared to distinguish the attitudes and the residential arrangements and maternal caretaking practices once the child was born (Kornfein, Weisner, and Martin, 1977).

Some appreciation of the range of differences in the families we are studying, particularly in areas influential in the child's development, can be gleaned from the précis of family styles that follow. Each narrative summary rests on systematic empirical quantitative analyses of value systems, social and physical characteristics of the household, parental aspirations and expectations for the child, observations and other documentation of caretaking practices, assessment of support systems available to the mother, analyses of the stability in family residence, family life-style, and significant persons in the household during the child's first year of life (Eiduson *et al.*, 1973; Eiduson, 1974; Weisner *et al.*, 1976; Eiduson and Weisner, 1977; Eiduson, 1978).

In the following summaries, each of these aspects is not treated for each family unit in didactic fashion; rather, we have emphasized what we found are the most significant characteristics in order to show how the family lives, what it values, what is the correspondence between values and practices, and what it provides for its members who call it "family."

Social-Contract or Unmarried-Couple Families

Some alternatives express interest in developing strong humanistically oriented relationships between individuals where commitment to a husband-wife type of relationship rests on emotional rather than legal bonds. "A relationship should endure only so long as it meets the emotional needs of the individuals involved" was frequently expressed during the heyday of the counterculture and became the tenet on which some young families based their conjugal experience. The notion of commitment to one another became a common expression—this was paradoxical and contrasted outwardly, at least with the existential immediacy in gratification and the lack of long-range planning that were also valued by these same young people (Coffin, 1972; Macklin, 1972, 1977). The word *commitment* was also in sharp contrast to the motivation for what turned out to be an experiment often not so different from the cohabitation living experiments that were tests of how acceptable couples were to each other, of whether or not they could evolve a viable living arrangement that permitted the freedom they sought within an emotionally powerful relationship.

In many ways, the social (rather than legal) contract families were identified most with the alternative philosophy. They wanted to break down stereotyped sex-role family arrangements, abandon the focus on possessions and materialistic arrangements, and strive to maximize individual needs in a loosely structured but psychologically important relationship. Some had been so sensitive to the unhappiness of parents through divorce and separation experiences that they were "running scared." Though some women particularly expressed a desire for marriage to protect the status of their offspring, they were able to tolerate the uncertainty in this arrangement because the quality of interactions made them feel both secure and free at the same time (Kornfein *et al.*, 1977).

Within the unmarried-couple families were a subset who could not wed because of legalistic complications. While committed to each other emotionally, many of these would have preferred being married. Unlike the couples who were expressing their views about the "pure" *raison d'être* for a relationship, they were less able to assume personal responsibility for their lives; rather, they were prohibited from being able to make a free choice. Yet, interestingly, these differences in overt motivations for being unmarried did not significantly influence their identification with the values found in committed unmarried couples. Social-contract couples put a premium on a more egalitarian role for both marital parents, were interested in achievement for its own sake rather than for materialistic goals, and valued interpersonal relationships and the intuitive subjective bases for decision making.

In everyday matters, however, the more emotionally "committeds" were more prone to have mother and child use the father's last name, put two names on the hallway mailbox, and were more open and unguarded in telling their parents about their legal status. The "circumstantial" families tended to bow more to circumstances in their lives; yet a man might accept a wife's arguing to become legally married to please her or to protect a child from "bastardy" but just as often, one might ignore parental pressures, income tax, and inheritance issues. That such issues had times and places was evident in the fact that in this unwed-family life-style category, there were eight changes in the direction of traditional marriages during the first 12 months of the child's life. In this same period, 15 couples split up, so that the unstructured arrangement seemed more conducive to change than, for example, that of the traditional marrieds, who were the most stable of our groups so far as residence or life-style shifts.

Parenthetically, the traditionally marrieds in nuclear units were significantly more similar in many aspects of their value affiliations to the traditionally marrieds who were also in living groups than they were to any of the other alternative populations (Weisner et al., 1976).

The family life of all the unwed couples is probably the most unscheduled, loose, and day-to-day of any group. Work-determined routine exists to some extent, but the Puritan work ethic has been supplanted by a more catch-as-catch-can existence, with seasonal pickup jobs demanding minimal skills. Income from such work is usually supplemented by welfare to the mother—another factor encouraging nonlegal relationships. Constrained by an infant and yet valuing equal family roles for both man and woman in the relationship, women rationalize welfare as their contribution to the household. Families live in apartments, small houses, some casually decorated, some carefully "funky." Stereos, records, books, pictures are common to all.

How casual are other life aspects? Child caretaking is usually intense and even compelling around the clock, since infants are seldom left with friends, baby-sitters, or grandparents. They go to school, to the movies, and to evening parties, accompanying parents everywhere. Breastfeeding is long-lasting, both for child-rearing and economical reasons. Fathers enjoy their "distributed parenting" roles, although they do not participate as much as they stated that they planned to in the prebirth interviews (Eiduson et al., 1973). Mothers feel comfortable asking for paternal assistance. Fathers also feel comfortable about telling mothers why they do not want to help at times. In one household, a father tried to discourage a second child, although he had played a large part in rearing the first. When it arrived, he refused to care for it, saying it was the mother's. Later on, however, he became less adamant in attitudes and responsibility toward the appealing little girl.

In this open family setting, the child sees all and may know all (cf. Ariès, 1965). He or she is expected to conform to the family model looking out for self at times and yet being a part of the larger family unit. Family routines shake down around what the family has to do to make a living, see friends, and participate in their music, community, and natural food activities. While a youngster's schedule may be loose and subject to change, parents so quickly incorporate a child into their unit, almost as themselves, that the child invariably gets the quantity and quality of attention that the parent wants for his or her own self.

Single-Mother Households

The single-mother household in our study reflects the change in attitude among these contemporary single Caucasian mothers, who keep their children and raise them without the guilt associated with having an out-of-wedlock child seen in previous generations. Our population of 50 women in this group originally identified themselves glibly with the characterization of the modern liberated woman. However, as their motivations around having a child were explored, we found more intragroup variability than anticipated (Kornfein *et al.*, 1977).

The Single Mother Who Is a Professional. Some professionally trained, capable, and ambitious women desired an expression common in the womens' movement, "the total female experience." This meant fulfilling goals involving motherhood as well as career. Thus, they proceeded to get pregnant, choosing often a biologically preferred man as the father. For these women, marriage was not in the immediate offing, though not ruled out. Pioneering in their orientation, buoyed up by educated talents and tested competencies in the world of work, they perceived a child as enhancing an already full and vital life. They consciously planned to start a new family unit, to "build a nest." Many expressed a preference for a baby girl, and they did have a slightly higher proportion of girls than did the other groups, although the number in the total population was about equally divided between sexes.

Most women approach motherhood for the first time with trepidation and anxiety; these women were not different, but they did anticipate the experience with more optimism and feeling of control over their fates. However, like the traditional mothers, almost all had hospital deliveries, feeling that supports were not sufficient for a home delivery. Although some were determinedly identified with some planks in the alternative philosophy, such as sex-role egalitarianism and humanism, their values were traditional in other respects: they valued material gains, were achievement-oriented, and had long-range plans and goals for their lives. They were self-sufficient economically and psychologically and became active and creative in developing and maintaining social relations that would provide enough personal contact and support to keep them engaged in the community (Kaplan, 1978).

Most of the women live by themselves in apartments that reflect sophisticated tastes and interests. They have friends through both professional and neighborhood contacts. A few are active members in professional or social groups, but they find that work and child caretaking occupy their lives, especially when the children are small. However, they are conscious of making time for social contacts so they can go out on weekends or in the evenings. Once the child is older, with regular caretaking activities, the women become conscious of being lonely. They desire more male companionship and want to get married, despite the very active pace of their lives. Some have tried to interest the biological father once the child was born, but without much success. At one year, out of the entire group of 50 single mothers, only two had married the biological fathers, and two more were legally married.

The child's life is like the mother's: regular caretaking arrangements, usually in private homes most of the day until the mother picks up the child en route home. Relatives occasionally assist, as when children are ill, or the day-care parent is away. Once home, mother and child are invariably together. Enormous emotional attachment and dependency grows up between mother and child, despite the fact that both spend large parts of their days with others. Weekends they usually do errands and visit relatives, go to the park or to the mother's friends, some of whom also have a child.

Single Mother Managing on Her Own. Two other groups of women who decided to head family units found the antiauthority stance of the alternative movement an ego-syntonic rationale for keeping their baby. Both groups became pregnant without planning to do so and then opted to become like the single women who were setting up new families and ostensibly managing on their own. One type of woman tried to manage a family on her own resources; the other moved in with her own parents.

The single mother who lives alone and who has less educational and financial resources than the professional single mother leads an existence that appears haphazard. Less goal-directed and future oriented than her professional sister, she usually moves among work, welfare, and school in order to make ends meet, develop some skills, and make an interesting life for herself. This seeming variety of activities is matched by variety in living arrangements (Klein, 1973). Sometimes she shares quarters with female friends—those with small children find each other—or a few with male friends. Sometimes the men are boy friends, who may or may not be interested in or attentive to the child. While the mothers welcome the assistance or the response of their man to their child, they often are so psychologically needy that they try to fill their lives with adult relationships in which the child plays a secondary role.

As with activities and relationships, these mothers are usually dependent economically. They cycle in and out of welfare, which they supplement by irregular employment, so sharing expenses with others is usually a necessity. Their homes reflect their marginal existence and attest to the uncertainty of goals and the superficiality of involvement that characterize these years for the mother of a young child. Unlike the career-oriented single mothers above who are so much like conventional parents in values, these women are less future-oriented, adapting to the circumstances in which they find themselves, less identified with the aims of the women's movement, and more drawn to the alternatives' interests in self-help and participatory groups. The self-reflective intuitive philosophies, usually newly introduced to them, became attractive ways of establishing social bonds.

Some of the self-poised and experimentally oriented women travel from one locale to another, occasionally dropping in on their own parents. A few (three during the child's infancy) went into living groups, which hold promise of social, economic, and childcare supports. A child is important to these women as one tangible evidence of permanency in their lives, but the child is likely to be regarded with ambivalence when his or her needs interfere with the parent's.

These women, like the career-oriented single mothers, want to marry. Their dependency and youthfulness appeal to some men, and while subsequent relationships may not be without conflict, they tend to settle into a more stable and regular routine. Children respond to the security this brings and, in their early years, quickly adapt to "Daddy," and a traditional household emerges.

Their limited skills foster homemaker roles for these single mothers. Those who remain alone live in lower-middle-class neighborhoods, in line with their limited work incomes. The women may do sales or business work or may get training as paraprofessionals, but they do not appear strongly career-directed. The children are cared for at home or in neighborhood play groups before moving into preschool settings. At age 3, they may stay away from home most of the day, going from preschool to a neighborhood day-care center.

Single Mother Who Lives with Her Family. Only a few single mothers in our population reside with their own parents. Most of the single mothers have been independent of their families of origin prior to becoming pregnant, so do not look to their own parents for support once they have a child. However, a few of our single mothers became pregnant while living at home or close to home and so returned during these trying times.

This subset of single mothers is—in personality, psychological resources, and life-style—reminiscent of the unwed young mother, whose fate was often identified with economic and emotional deprivation (Gershenson, 1972; Furstenberg, 1976). In our sample, these women proved young and naive, had limited educational and vocational competencies, and had become psychologically and economically dependent on their families of origin.

This mother-child family unit usually becomes integrated more or less with the grandparent family. While the mother takes care of the baby and often shares sleeping quarters with regard to eating and other family functions, in our households, mother and baby become merely two other family members. The mother is usually treated much as she was when part of the family originally, motherhood giving her no special status. The baby, however, takes the role of youngest and is enjoyed by all. Few of these mothers have trouble using all the affection and assistance with the baby that the family gives; only "advice" causes problems. Advice about the child infantilizes the mother in the one role in which she has overtly grown up and matured, and it is a frequent cause of conflict.

The sensitivity of the grandparents to the mother's circumstances usually distinguishes the family units that are comfortable homes for single mothers and babies from those that prove temporary. If the grandmother's own marital state is unsettled, her daughter and child can prove to be particularly burdensome psychologically.

Economically, mainly through welfare, the mothers contribute to board and room; some get part-time jobs, but since few want to release parenting to grandparents or sisters completely, they take care of the child during the daytime hours. The mothers remain eager to have personal and social lives of their own. "Dating" behaviors usually restimulate the grandparents' anxieties and angers over earlier acting-out behaviors of their daughters, and the resulting tension can force a mother to move out of town, leaving the child with the grandparents. The grandparents are ambivalent about being given total responsibility, despite their love for the grandchild. They seem to see the baby as continuing evidence of their daughter's instability, immaturity, and rebellion, and they resent having to assume the consequences.

A child left with its grandparents is usually the only child in the household. He or she is brought up as an only child, of older "parents," and so frequently is fawned over and given many advantages. The stay with the grandparents is, however, usually temporary, ranging from a few weeks to two years, in one case, because of the mobility, the impulsiveness, and the rebelliousness of the young mother.

Such behaviors on the mother's part might have occurred without the support given by the youth rebellion of the 1960s, but the ease of finding confrères who share their views gives new credence to the developmental battle against parents. Their "here-and-now" perspective also reflects impulsiveness and need for ready gratification. So many of these women seem strongly aligned with alternative positions on the surface, but in their interest in possessions, in their desire for the goods of achievement, however unrealistic in terms of their skills and competencies—in these regards, they are like all the single mothers, who have strong leanings toward traditional values.

Communes

The 50 participants in living groups, intentional communes, or more appropriately today, communities, are variable along a number of dimensions that prove to be relevant for their ways of functioning as a family. They can be small or large; rural or urban; creedal, that is, having an ideological commitment; or merely social or proximal domestic units. In size, extremes reside in groups that are triadic, with two like-sex partners mated to one spouse; or groups can be the combinations of kin and nonkin "extended families" of hundreds seen in the more popular religious groups.

We have found it useful to distinguish between the creedal or religious groups and the domestic groups (Berger, 1971; Blois, 1971; Cavan, 1971). The former are based mainly upon either Eastern philosophies and culture or the new "Jesus Movement," which seeks a new way to live out traditional Judaeo-Christian convictions. A smaller number of creedal groups are derived from allegiance to a charismatic leader with a more-or-less formalized philosophical system. The domestic group, by contrast, is a less formally organized unit, often formed spontaneously around shared crafts, political convictions, or identification with the natural rural environment or the small contemporary farm. In the latter, people who like each other decide to live in a small household or in close proximity and to share some quarters and tasks and some social and domestic activities, often child care. Some of the implications of these differences in family composition, structure, roles, and responsibilities are evident in the following description of their functioning (Minturn and Lambert, 1964; Greenfield, 1974; Weisner et al., 1976).

The Christian Religion As a Family Group. The fundamentalist Christian groups have found a new impetus from young people who joined them during the late 1960s and early 1970s in order to find a way back from a drug-ridden, unstructured, and meaningless counterculture life. The Christian groups that attracted them had a long and usually uneventful history of small-time evangelizing and proselytizing. However, these groups were energized by these young people who sought ways of giving new meaning to their lives in pledging themselves to Jesus.

Christian groups to which our participants were attracted were spotted up and down the California coast. Though the specifics of their lives were somewhat determined by the leadership, locale, and history of the individual group, the form of family life was common to all. Members lived in individual nuclear family units, usually apartments—or small houses, if married and with a child—so there was some private family existence even though many of its details were determined by the common practices of the group. When people arose, their time of prayer, the place and nature of their activities, who they worked with and where, and what they ate were determined by the group. However, our families almost invariably ate noontime meals, napped and slept, dressed and played in their tiny private quarters. They were interviewed and observed at periods of home observations in these quarters, even though their participation could not have proceeded had the elders not agreed to it. When members from one group within a sect would come to project offices in Los Angeles, arrangements for housing them would be made with elders in the local groups.

Children have a special place when reared in these Christian homes. They are the future evangelists, tomorrow's truth sellers. Their own activities, growth, and development are governed with this in mind. They pledge themselves to the Lord, usually by age 6, but prior to this time each has regularly attended and participated in prayer service. Being part of a group helps make palatable the expectations that they will assume some assigned responsibilities from the time they are 2 or 3 years old. Similarly, prayer rituals at meals, service attendance, and the like are proscribed behaviors.

Parents regard the Lord's will as dictating the child's personality and his response to parenting practices. It is then up to the parent to work with the child to make him able to be responsive to the Lord's teachings. Spanking is a common disciplining tool in line with the Biblical "spare the rod and spoil the child"; invariably it is followed by hugging to show the child that the real intent is punishing the child, with the parent as God's representative of love and care. It is not uncommon for a child to be lightly spanked three or four times during an assessment in our offices—with the parent showing none of the guilt that a less righteously and ideologically supported parent would show.

The children, like the parents, live modestly, wear cast-offs from older children in the group, and play with brothers or sisters of other children in the group in the fields or in an informally developed play area. They are also expected to be responsive to whoever takes care of them—for from their first year, their own parents were assigned to tasks that took them away from home, while other women were caretakers. Child caretakers change from time to time, and the child is expected to adjust regardless. The caretaker has complete responsibility but knows where the parent is if necessary. A preschooler might report to his mother that he has been hurt on a wire fence and that an older boy has taken care to bandage the wound. The mother would assume that an appropriate intervention has been taken, as indeed it usually has.

As the parent accepts certain callings of the Lord, so the children assume that their lives will be regulated from the outside. Even when the children go to public school (since church schools are not always viable alternatives), the patterns of scheduled activities at home are also adhered to before and after school. The times for fun or play, for school, for meals, and for prayers are set. Developmental patterns are self-defined. "Proper" behaviors are sex-stereotypic, with both adults and children in traditional roles (Weisner et al., 1976).

Family life, then, is rigorous, scheduled, and rationalized by goals larger and more important than any individual goals, or any individual. Prayers are said for the errant husband who has not yet been brought into the Lord's fold; a new relationship might be well thought-of for a single woman if she needs help in controlling a child—for example, she might be urged to look for a husband. Similarly, with pregnancies, number of children, rebellious or wayward teenagers. Like their parents, the children learn what is expected of them early. Parents have few compunctions about bringing their children into line, for the group condones practices that might elsewhere raise doubts. Life, while solemn and directed by powerful purposes, is invariably real, harsh, and so a test of commitment of oneself to large goals.

The Eastern Religions As a Family Group. Doctrines of various Eastern religions differ, rituals differ, and relationships and involvement with the outside community differ. However, the family structure within each of those being studied have certain common components (cf. Ornstein, 1973; Judah, 1974).

Like the Christian religious family groupings, the Eastern religious families have a family life that is private and outside the group experience as well as a family domain that is group-shared.

The family structure for a child is usually nuclear—father, mother, child(ren)—but families do separate, with one parent then remaining in the group. In most communities, nuclear families have separate quarters or a private space—an apartment, a duplex, or a bedroom upon occasion. Quarters are austere but beautiful. Furnishings are minimal, with floor pillows replacing furniture; walls have religious colored paintings. A feeling of calmness, dignity, and care pervades. Meals are generally eaten with the group—for like the Christian groups, these have ritualistic features, sometimes dietary restrictions or conventions about prayer or interpersonal communication. The children are introduced to these early, for like their parents, they participate in both the private and the group-shared family practices.

The parents are usually expected to nurture their own young, with supports from group members, who may baby-sit, run a preschool program, or carry out those activities that make it possible for the mother to play a role in the common activities of the group. In some families, the mothers are heavily involved in community functions; in others, more minimally so when the children are very young. Fathers who are more traditionally in novice or priestlike roles assume both economic and religious tasks that make their family involvements very much like those seen in the traditional middle-class families; that is, they are present for meals and occasionally also in the evenings, but they are expected to be mainly occupied with the work that makes the religious family go. Some religious groups exist economically through providing services to the community that have nothing to do with their religious mission—at shoe or grocery stores, manufacturing natural foods, etc. While both men and women can participate, assignments by group administrative leaders see that men participate regularly. Their spiritual activities are then fit into mornings and evenings.

Parents then support the larger group by some assigned tasks and in turn are freed from the need of individual supports: these are provided by the group. They have food, shelter, and friends who share spiritual and personal commitments and who provide all manner of personal assistance and resources. Depending on the extent to which a religious group is closed or has involvement in the larger community, a close friendship can also exist outside the family group. Some religious members are sent to school in the community to obtain competencies and skills that the group needs, to get some of their own space, to relieve the crowding within the group confines, and to serve as a representative of the religion in the larger outside. The children are similarly given some opportunity to partake of what the outside offers: public schools, companions, family friends and relatives, grandparents. How far the personal vista extends and the way it shapes, influences, and is integrated into the family's total life usually depends on where the religion is going and what are its overall goals. Just as American versions of an Eastern religion differ from the model they emulate, so family life-style is less rigid than in the form of the family historically identified with the ideology.

Eastern religions offer few definite principles to guide young parents in child rearing (Kornfein, 1975). Food regimens define kind of diet and times of eating but not necessarily eating habits. Even in those religions in which the children are isolated alone with the mother for days following the birth in order to protect the newborn from unwanted influences, creedal rules do not cover the later, less ritualistic aspects of mother-child interactions or child-rearing practices in regard to discipline or toilet training. At many intervention points, the parents make their own decisions, and thus a child sees a model in which the belief system provides the general outlines for life, and the parent gives it his interpretation.

A visitor to a group of children in a religious group who are easily distinguished by dress and aspects of physical appearance may find them playing with wagons and tricycles, digging in the mud, or squabbling over dolls.

As the children get older—and in some groups, this can be at age 2—special schooling oriented toward religious beliefs but also incorporating sensorily stimulating and artistic materials is started. The spontaneity and creativity of the young, impulsive child are in some creedal belief systems highly valued; the child is unfettered by most conventions in these orientations, except for a general religious upbringing. In these cases, the roles that children will assume as young adults begin to take shape in very early years. Invariably, early orientation toward religious roles follows sex-stereotyped lines.

Where the child-rearing conventions are flexible and not predetermined, the parents depend on discussion with each other or with the spiritual leader for questions about troublesome children and divergent adult preferences.

The Leader-Directed Communal Family. Religious family groups are in a sense leader-directed; however, there are also family groups in which the leader is not a religious but a charismatic figure. Leader-directed families are usually distinguished by their devotion to specified goals and aims that serve as ideals and by explicit group adherence to ways of achieving these ideals. The family's sense of unity comes from respect, admiration, and love for the family leader and a willingness to be part of the community who adopt his philosophy. The authority orientation of these families is similar to that found in charismatic-leader families, and in fact, some of these groups consider themselves religious, meaning devout, with members as devotees, but their allegiance is to a belief system that is not structured around the concept of God. Instead, tenets of the leader show "The Way," with a search for self and for one's role in the social scheme of life often being the focus. This is the reason that these groups are sometimes referred to as psychotherapeutic cults (Singer, 1978).

The group context in a leader group provides a family organization that has social relationships and interactions and a hierarchy for roles within the group, as well as some organized or considered sets of relationships with the outside community. "Someone from the outside world is here" is sometimes heard in one group as a nongroup member visits, but even more casual interactions with outsiders usually have specified goals and constraints.

The leaders of community family groups have been called charismatic to stress the power and attraction for followers. In other characteristics—appearance, voice, demeanor, intellect, personality, and past or current achievements—these leaders have few common denominators. All, however, address with some more-or-less conscious awareness the universal needs for feelings of being different (i.e., superior or elitist), or searching for the road to inner peace through helping to achieve some social goals or ideals that the group desires. In some, it is to produce a special breed of children, who will be future leaders; in others, to develop great emotional sensitivities or creative powers to elucidate the unusual psychic capabilities of man. Consistency, integrity, or rational synthesis is not sought in the philosophy, nor do discrepancies in logic trouble family members. Instead, if there is coherence between life-style and belief systems, and/or if the philosphy effectively rationalizes family practices, its tenets are usually not questioned.

The development of the personal, but shared, philosophical views encourages the development and practice of cult-like rituals. Practices that can emerge may be polyandrous practices, in which a few powerful men in the group sire offspring with various women, or, practices in which parents allow children to be separated from them and placed with group or infant-school caretakers who will see that the child's special powers or "energies" are used to group advantage.

In a group that believes that children and adults are individuals and, regardless of age, should be sensitized to their needs for sensuous expression, group activities may involve children or teenagers. Sometimes sexual activities develop naturally in the course of relationships and then are later rationalized, but as often, a group's philosophical position anticipates and thus fosters these behaviors. Ideology gives significance to behaviors and to the members ascribing to the ideology. This is the reason that the information base or cognitive background is seen as the strength of the cultist groups (Singer, 1978).

Children who grow up in a leader group are often reared in rather radical ways that again have a stated correspondence with the ideology that the adults live by. A 24-hour nursery may focus on enhancing cognitive skills or hedonistic pleasures through overstimulation or on building children who do not distinguish kin from nonkin parents in the zeal to build for an extended family rather than a two-parent bond. In these, the ideology of the group more-or-less directs child-rearing philosophy and practice, but these may turn out to be experimental, erratic, and radical. Since there are few tested precursors for child-caretaking arrangements and few adults trained as educators or group day-care workers, it is a challenge for novice caretakers to provide the care and attention that the usual small family model provides for a child. There is not one but many children to attend to and to relate to. There are also decisions to be made constantly about the conflicting needs of children and adults, as we well know now, observing the struggle of restructured families to maintain a cohesive and workable structure that resonates to the needs of both children and adults.

For the children, the group experience is thought to provide companionship, allies, models for learning, and the social competencies that come with having to find ways to live with many other children and various caretakers. Children are usually expected to find their own ways of using the social and cognitive environment. There are examples of strong bonds with parents who direct, encourage, and single out their children for attention. Parents may desire to do this, but if attention is too intense, it is sometimes discouraged as making the child (and the adult) too dependent on each other. While not openly avowing that bonding with the parent may diffuse the child's alliance with the leader and the group, this is apparently one of the aims.

Children usually eat, sleep, toilet, play, and are disciplined and taught in the presence of each other, so there is usually a structure about what goes on in regard to children—even in those family groups that strive for casualness, unstructured interactions, and spontaneous interchanges. As with brothers and sisters in one very large biological family, the children learn to make their ways with each other, and have personal attachments, rivalries, and toy exchanges.

The Family in the Semirural Domestic Group. The desire to return to the land and a more simple, traditional rural existence brings this family group to the small, almost hidden community in remote areas (Gross, 1968; Diamond, 1971; Steiner and Maran, 1971). Their simple houses show a mixture of minimum conveniences and sophisticated books and electronic equipment. The family has usually gathered a group of interested friends; has decided to live in proximity to them so they can share companionship, some meals, and some activities; and has also usually demanded at the same time the kind of privacy that can be obtained only in the private dwelling.

A family, then, is surrounded by people who share interests

and tastes, who like to exchange with close neighbors their propensity for working out those aspects of life for themselves that are usually taken for granted—food from market shelves, inside toilets, large houses, and comfortable furniture. The process of building up one's own wherewithal for existence becomes the sum and substance around which family life proceeds. As in the farm family in the 1800s, there is little time or energy left for less mundane, more sophisticated activities. Pleasures include simple respites from hard work, often drudgery, but the greatest gratifications rest with the fact that the family is constantly renewing its vitality and viability by its own hands.

This family unit, then, creates what it needs mainly with the supports available from carefully selected friends in like circumstances. The economics of this living style are usually a big issue—welfare, clipped coupons, and occasional outside work of a physical and unskilled nature (carpentry or wall papering). Resources buy the seeds for the garden and the small farm animals and keep the old car or van running (Mungo, 1970; Roberts, 1971; Kinkade, 1972). The children have the freedom that has always been provided by farms, busy parents, and an almost exclusive concern with the basics of life. They are around constantly, participate as they can in the parents' activities, and play in and out of the house with makeshift toys and spontaneously thought-up games. Their lives are regulated less by their individual needs than by their family's needs—an early-to-rise, early-to-bed household trains children accordingly. Their food and toileting practices and discipline have the same up-front, immediate character. However, it should be noted that the children's upbringing reflects their parents' sense of inner discipline which is apparent in their patterns of regular work, efficiency, conscientiousness, and determination.

Children seek out companions from the intimate circle of family friends residing close by. An occasional playmate may be found in an informal nursery group prior to nursery school, where usually a more middle-class structure prevails. School initiates the combination of home freedom and outside-family structure for the child; both serve usually as fine complements to each other. The parents who have had children always around also enjoy the freedom. Trips to the laundromat, the natural food stores, and the occasional movie are treats—as are candy, Big Macs, toys, and visits from grandparents.

There is little ceremony in these families. Informal parties and volley-ball games are as frequent and as taken for granted as the slaughtering of animals for food or the continual building, improvising, and enlarging that goes on. Socializing has a useful supportive quality and permits the kind of interpersonal examination and discussions that are meaningful to these modern small-vegetable-garden farmers but were not part of the traditional American rural life style. As these families find less need to prove their self-sufficiency in this self-selected life pattern, they venture into the nearby small towns, go to classes and lectures, see friends, shop, and begin to move increasingly toward the life of the small urbanized community. Their child does too.

The Urban Domestic Family. The child in this family shares with the child in the rural domestic group the kind of freedom that comes from being present and constantly with its parents and their friends (Eiduson, 1978). Life has a rather informal, relaxed quality, with parents and children living in as easy and flexible a way as possible. When the parents work away from home during all or part of the day, the child's day is shaped by the caretaking arrangements that the parents have made: day care, neighborhood nursery group, or baby-sitter. The parents place minimal demands on themselves, trying mainly to do the things they like to do, seeing the people they like to see, and interacting openly and without ceremony. The parents' demands on their children for behavior or conformity are light. The parents see themselves as supportive, meeting minimal needs without pressure, and the children adapt accordingly. They relate to their parents in a casual way, seeing other adults and children as important persons in their environment. Child rearing has a casual, spontaneous, sometimes impulsive quality about it—in contrast to the planned, scheduled milieu found in many traditional homes. Doctors are seen as necessary, schools used when convenient, friends visited spontaneously.

Some family units share a large house, with parents and child sharing small bedrooms. Others divide rooms more systematically, so like-aged and unrelated children sleep together and separate from their parents and the other adults. Sharing is usually a major philosophic tenet in these arrangements and can be the most planned aspect of the family—and the most conflictful. The economics are shared, as are decision-making, homemaking jobs, etc.

Child caretaking is one area where the concept of sharing leads to freeing the mothers for other activities. This can be done with forethought and care and become a major concern of the group—at least of the women—or it can vary on a daily or weekly basis. Many women have been motivated toward the living-group household because of the opportunity offered to vary traditional roles and permit more personally gratifying educational, social, and vocational experiences.

By contrast to the religious and charismatic-leader groups, there is less stability in these last two domestic communities. All these communities taken as a group lost more memberships (one-third) than did the other three life-styles; however, the creedal groups were more stable than those that were more informal. This finding is supported by studies of the viability of communes in the heyday of the countercultures, which pointed out that groups that had an explicit goal or philosophy, a charismatic leader, and a formal organization and that were economically viable were most likely to survive (Eiduson et al., 1973).

Traditional Couples

Our comparison group, the traditional marrieds, are a random sample group. We found, probably not surprisingly, that the contemporary young, conventional, two-parent, traditional, Caucasian middle-class household embraces a range of values, family practices, and beliefs.

Conventional Two-Parent Nuclear Families. For most of the legally marrieds, the two-parent nuclear family, legally attested and tied into an established framework of kin and neighbor relationships, is still the ideal family model. The wife is expected to maintain the domestic sphere, while the man is designated as the material provider, but interestingly, in this group are found the largest proportion of part-time or full-time working women. They have received more formal education as a group and more often have had careers prior to childbirth. More than half of the legally married sample have been with their mate five years or more, and 90% three years or more. This is longer than the typical relationship in social contracts and, of course, much longer than that between single mothers and their men. They are also far less likely to have been divorced; for 90% this is a first marriage. The father is almost always employed, and family income is at the national urban average, higher than in any other family life-style. Eighty percent live in houses, the rest in apartments. Furnishings and size of house vary a great deal, depending on the current income of the family.

The traditional units show the range of behaviors, attitudes, and values that has been identified by other investigators' studies of middle-class families (Sears *et al.*, 1957; Sears *et al.*, 1965; Baumrind, 1971). However, as a group, we found these families very different from the nontraditional families. Among areas in which they differed were in aspirations for their child, the traditional being ambitious for their child to go to college and achieve fame, money, and status. In line with this perspective, they themselves strive for the same goals, are aligned more with a Puritan ethic in attitudes toward work, and are comfortable setting long-range goals for themselves. They strive for objectivity in problem solving and value rationality and intellect as modes of making decisions. In regard to their own health, welfare, and education, they use only existing institutions. However, interestingly many of the women are planning to return to careers when they are able, have part-time commitments because they are conscious of the importance of keeping up skills, and consider that a combination of home and outside interests and activities would be feasible and stimulating.

Nonconventional Two-Parent Nuclear Families. The "nonconventional" traditional family is the family that to all outward appearances is a traditionally married, two-parent nuclear family but in its values and within-family practices is alternative. These are families in which the work roles of the parents may be reversed, or they may be couples whose life-style departures from the mainstream go in the direction of being so affiliated with one of the cults—like the Eastern religious authority-directed vegetarian sects—that this allegiance directs much of their lives, even though in outward appearance they seem conventional by working in mainstream occupations and living in middle-class neighborhoods.

These families have sometimes been called *marginals* or persons with one foot in two cultures, for they appear to be populations who are on the borderline or the cutting edge of change. They experiment with change in various areas of their lives and seem to be resonating to pressures in society that promote testing out practices that are deviant in certain ways. These practices are occurring within sufficiently conventional boundaries so that the front they maintain to the world gives little suggestion of their propensity for alternativeness. They are a legally recognized entity, live in conventional areas, send their children to conventional schools, and are accepted by neighbors as mainstream members but operate in some specific and significant ways as very nonconventional.

It is also tempting to consider the unmarried-couple families and the single mothers "marginal." Yet the legal and conventional front that network traditionals maintain gives them a privacy and a camouflage that seems to permit a wider range of within-family behaviors than the other unconventionals have.

In these families, child-rearing practices follow the styles of living that the families evolve. In the case of the father who becomes the "house-husband," he takes over the major child-rearing activities, although the mother may remain the decision maker in child care (Eiduson and Alexander, 1978). In the case of affiliates of the contemporary sects, we find that they operate with their own mores and nutritional and religious practices. These, however, are so circumscribed or so effectively interdigitated with the conventions of the life-style of the dominant American culture that their unusual beliefs or behaviors are only upon occasion conflictful.

The life-style of the young children in these families is by and large that of the child in the traditional family. Both the outward social conventionality of the family and the child's own contacts and activities outside the family make for a fairly typical "middle-class" experience, especially since many of these people live in conventional residences and neighborhoods. The child, like the parents, is pretty indistinguishable from friends and age-mates. The differences that make the appositives *traditional* and *network* good characterizations of the family do not seem to generate obvious problems for the young child; the effects in later life remain to be seen.

ADULTS ATTRACTED TO ALTERNATIVE FAMILY STYLES

To understand the motivations for entering into families, we examined a large number of demographic and social background variables, such as place of birth, birth order, years of education completed, present perception of past relationships with parents and siblings, and previous drug history (Cohen and Eiduson, 1975). In general, our data supported findings by other investigators, who had shown that young people who had opted for variations from the nuclear family were, in the main, not different from their peers of like social classes.

First, some general population characteristics:

1. Mothers in the study were comprised of a group of 208 women whose mean age was 25.4 ± 3.8 with a range of 18-33. The fathers' average age was 27.9 ± 4.9, range 19-49.
2. The greatest concentration (well over half of the mothers and fathers) were born in the Pacific states, with the Northeast and the Mid-Atlantic states contributing the second and third most sizable contingents.
3. At the time they entered the study, 50% of the families lived in Los Angeles and environs, with the remainder divided fairly evenly between San Diego, San Francisco, and the Northern California area.
4. Of the participants, 83% were currently residing in urban settings and 17% in semirural areas.
5. Both parents had an average of slightly more than three siblings and came from families with as many as seven children. The mothers were most often second oldest children, while half of the fathers were eldest children.
6. The socioeconomic status (SES) of the population in terms of the mother's family of origin showed that 17% were from upper-class families of origin, 18% upper-middle, 31% middle-class, 27% working-class, and 7% lower-class (although we had not planned to include this group in our population) (Cohen et al., 1975). The father's family of origin showed a similar SES distribution.
7. The SES of the parents at time of entry in the study was of

course very different, for the nontraditional families, by and large, had become downwardly mobile (i.e., were living at a level "below" that of their own parents), while the traditionals were more on a par with their previous backgrounds. With the nontraditionals, downward mobility was often a chosen departure in line with their distaste for materialism as a goal of life; their divestment of conventional kinds of possessions; and their more casual attitude to goal oriented employment.

8. The alternative group more frequently interrupted their education than did the traditionals, as would be expected from the literature on young adults attracted to the counterculture movement in the late 1960s. The total population of mothers reported completion of (mean) 13.9 ± 2.2 years of formal schooling; the fathers averaged 14.5 ± 2.6 years of schooling. However, women in emergent family variations are underrepresented in the group achieving college-level degrees, while traditional fathers have gone on an average one more year in college, but the differences between life styles in years of education completed are not significant.

In comparing the traditional parents to those living in one of the alternative life-styles, we found relatively few background variables and attitudes that seemed to differentiate the two groups; however, the factors that were significant contributed to our understanding of the motivations for seeking a family variant form:

1. The parents of alternative family participants had moved from one residence to another much more frequently than had those in the traditional life-styles.

2. Also the parents had remarried significantly more often.

3. Additionally self-perception related to earlier life experiences and collected retrospectively showed that as a group, alternatives viewed their early childhoods as unhappy, whereas most of the traditionally married parents perceived this period of life as happy.

4. There was a general tendency for the traditional mothers to have maintained better relationships with their own mothers during childhood and adolescence than had those women

who chose to live in an alternative life-style. Interview data obtained from the fathers in separate sessions showed the same trends. The mothers who were currently in living groups and in social-contract marriages appeared to have had the most difficulty with their own parents, especially in early adolescent years. Among the alternatives, the single mother seemed to have maintained the best relationship with her own parents in the growing-up years.

THE EFFECTS OF ALTERNATIVE SOCIALIZATION EXPERIENCES IN THE CHILD'S DEVELOPMENT

Studies of the physical, cognitive, and social-emotional growth of project children are compelling: they show that from birth through the first year of life, development is essentially normal (Zimmerman and Mickey, 1976; Eiduson, Zimmerman, and Bernstein, 1977). While there is some range in developmental scores, the mean for the group is average in both motor and mental areas. This was true when the children were tested at 8 months and again at 1 year. In fact, the correlation of test scores from the 8-month period to 1 year is positive and significant. Data in the cognitive area correspond to pediatric and nutritional status data and to socio-emotional assessment at 1 year, for in each of these areas, the level of development of the children, considered as a total group, is average.

Looking only at data collected during the first year, we explained our findings as the result of the convergence of infant socialization practices that our parents knew from personal experience. They were parenting as they had been parented (Eiduson and Weisner, 1977; Eiduson, 1978). For example, this first became clear during pregnancy, for despite their philosophical adherence to alternative values, the fact of pregnancy made an impact on these young parents who were in the main from middle-class and stable working-class backgrounds. They became concerned about the impact of their behaviors on the well-being of the child. For example, changes in drug use patterns and concern for nutrition changed dramatically by the third trimester of pregnancy (Eiduson et al., 1976). The amount of drug use during pregnancy decreased dramatically as compared to the amount of use during the years prior to pregnancy. The alternatives' drug intake was reduced to a far greater extent than was the traditional marrieds' because of their greater incidence of use prior to pregnancy. The traditional marrieds reduced their total drug intake to almost abstinence during pregnancy. (Incidentally the level of prepregnancy drug use in this comparison group was surprisingly high and made us appreciate the amount of drug use in a random sample of young parents in the 1970s.) While marijuana continued to be used among a fairly high number of mothers in the alternatives, the amphetamines and hallucinogenics are among other more powerful drugs that were practically eliminated. This trend toward less usage of drugs during pregnancy was also found to be true among the fathers.

The data are similar for food intake. In general, all the mothers were concerned about appropriate nutrition for the health of the child. No significant differences in health or nutritional status during pregnancy were discovered between alternatives and traditional mothers or among various alternative groups.

Organized medical services were used by nearly all the mothers in prenatal care regardless of life-style and regardless of their attitudes toward established institutions in the society. While the majority of both traditional and alternative groups planned to use hospital facilities, there was a significant difference between the two groups in their plans for home or hospital delivery. The alternatives more often planned to have the delivery in their home with significant people in their lives attending the birth. Prenatal care was practically universal, and 98% were involved in some sort of preparation for childbirth, as Lamaze or Bradley courses. Families went to clinics or to physicians known to be sympathetic to home births or nonconventional families, but there was only very occasional use of nonmedical interveners, such as naturopaths, herbalists, or nurse practitioners (Eiduson, Bernstein, and Preston, in press).

Of our babies, 20% were born in the home and 80% in the hospital. There was a significant difference in the groups in which this occurred; 34% of social-contract and 30% of living-group parents gave birth in the home, compared to 6% of the single mothers and 8% of the traditionally married families. We should not have been too surprised at the trend of these findings—knowing that when anxious or without relevant new experience, parents revert to practices with which they are familiar. Our surprise arose because these common practices in regard to caretaking and socializing the child occurred also throughout the first year despite the overtly heterogenous families in the sample, with their nontraditional life-styles, structures, family organization patterns, and values.

We interpreted this trend in findings as evidence that the baby with his biological and psychological needs were important determinants of what happened in the parenting process. We also found that the child-mother unit was the nuclear or basic unit in every single family, regardless of how oriented the family was toward multiple parenting, toward shared caretaking between mother and father, or how involved the parents were in non-child-oriented activities.

A variety of analytic studies were undertaken to see if we could identify some critical characteristics within the family that made for differences in the intellectual and socio-emotional development of children in the different family life-styles. Background variables such as mother's and father's years of school, their current SES, earnings, and the SES of their families of origins were explored, without finding a variable accounting for the major variance in Bayley scores, exploratory play scores, or attachment scores. Among current environmental features—number of people, number of adults, number of other children, whether the project child was a first or second child, whether the family was rural or urban when the child was 6 months or 1 year old, the number of residence changes or life-style changes, and a host of particulars about the home suggesting its potential in terms of intellectual or social stimulation—did not turn up more than an occasional significant variable, certainly less than the number expected by chance (Eiduson and Weisner, 1977). For example, the percentage of available people who did some caretaking correlated significantly and positively with the mental score at 1 year. Also, if the children were taken care of by the father or other women in addition to the mother during our six-month observation, motor scores were enhanced. Breast-feeding also proved of interest: when the same variable was related to the eight-months Bayley Scales of Infant Development (Bayley, 1970), there was no relationship for the Mental Scale, but there was a marked relationship between six-month breast-feeding and the motor scale ($p = .07$). The direction remained the same: significantly more bottle-fed infants were in the lowest quartile, and significantly fewer breast-fed infants were at the same level.

Obviously, caretaking variables specifically had drawn our attention: in one study, in an effort to evaluate more thoroughly the implications of relationships available to a child, parenting units were divided into those offering single-parent, two-parent and multiple-parent households. When these did not prove more powerful than the life-style in which a child was at birth, or at 6 months, or at 1 year, we then examined specific caretaking practices within families. Parent reports and home observations provided us with data on the proportion of time the mother spent with the child versus in other activities, the mother's employment, persons who took care of the child at night or at other times when the mother was away, and the frequency and duration of care by others—and 52 such caretaking variables that we hypothesized would tell us about the separations from the mother, the extent of the child's social interactions with others, and the consistency of caretaking. The caretaking practices studied singly produced non-significant correlations with Bayley scores (Eiduson, Zimmerman, and Bernstein, 1977). When examined together with multivariate techniques, three factors proved of interest for cognitive scores: the number of people in the family (positively correlated), the extent to which the mother was occupied with things other than the baby (negatively correlated), and the availability of other caretakers to the child (negatively correlated). These factors accounted for approximately 10% of the variance obtained on Bayley scores, and for none of the variance on the Strange Situation Test (Ainsworth and Wittig, 1969).

Therefore, it seemed as if the variations we have been seeing and reporting on family environments were not strong influences on early intellectual or early indices of socio-emotional development. The differences in the home milieus seemed not to be reflected in what the parents did with the infants, at least so far as caretaking, social exchanges, and relationships were concerned. The biological and psychological needs of the infant and the needs of the parent that are sensitized to the child become the salient determinants during year 1, whatever the varying ecological environments in which the family unit is embedded (Eiduson, 1978).

DEVELOPMENTAL TRENDS: THREE YEARS

Space does not permit presentation of the fuller results of the child's physical, nutritional, cognitive, and social-emotional development that have been systematically assessed from birth. However, mention of some data trends at year 3 is of interest. Almost half the sample has been seen for the three-year procedures, and for 91 we have fairly complete results (Eiduson, Zimmerman, and Bernstein, 1977).

Again, selecting only some findings in the cognitive areas: the total sample functions at the average level on both Stanford Binet (Form L-M) and the Peabody Picture Vocabulary Test, mean 103 and 102. Our children, as a group, seem typical of the U.S. school population in terms of mean scores, and the range of scores on our population is also typical, with standard deviations (15 and 16, respectively) similar to the standardization data (16) on both tests.

The Peabody Picture Vocabulary Test scores, measuring linguistic development, show some life-style differences, with single mothers scoring 6 points below traditionally marrieds, a difference that may prove to be significant as more complete Ns in each group are tested. Does this mean that the single parent converses or interacts with the child less? In contrast, is the child exposed less to two-way adult conversations? As we study life-style changes over time, we should have data on this very tantalizing hypothesis. The living-group children are in the middle, with those children of parents who are traditionally married *and* in a living group having the highest score among living-group children.

If these results on less than half the population are borne out on the total group, the consistency in cognitive performance is of very major importance, for as we have looked at "at-risk" children over time, we have also been studying what has been occurring in the families as their children have been getting a little older and less dependent on the parents. In the course of documenting socialization changes and changes within families, we have come across potential risk factors in some home settings characteristic of some of the problems common to families in today's society, such as: (1) marital discord and disruption; (2) separation of mother and child (and a break in early bonding patterns) after the child is 6 months of age, with the child put under the care of grandparents or nursery staff within the living group; (3) depression in the mother; (4) repeated residential changes; and (5) addictive patterns in a parent.

The frequency of such stressor situations in our population is not as yet established—nor do we know the extent to which what we are seeing differs from the statistics being reported in the society at large. What we can appreciate at this point is that alternative families are not immune from change, disruption, and problems. Although they have been entered into with the hope and expectations that they will offer the parents some options and potential for expression and development that the two-parent nuclear unit may not, apparently they are not free from the kinds of conflicts and tension that arise in any interpersonal relationships.

Why are signs of family disorganization now coming into view? Are the kind and amount of problem-laden situations we are beginning to see in some of our families common in American populations today but not so recognizable at these early stages in a child's life in the absence of the regular and periodic contacts and assessments we are fortunate to be able to continue? Perhaps. Our studies of our mainstream traditionally married group should let us know if indeed this is the case.

Despite the wide variety of life-style options originally represented in our families, certain elements in most of our family structures assume the traditional cast and seem universal—the consistency of the mother-child bond is one of these elements, for example. When there is a departure from or a breakdown of these, the child may be in trouble. We are documenting different and compensatory arrangements within the family under such circumstances in order to see how successful they are, in terms of the consequences for the child and for his parents.

The 1- to 3-year age period is gaining our attention at this time. How important is it for intellectual, cognitive, and socio-emotional growth? Developmental psychologists have pointed out the dearth of data relevant to this age period. Is it a particularly vulnerable age period? It may be, for in our population, important changes in the family and in the children seem to be occurring in these years: parents are disappointed with the unfulfilled promise of personal and social relationships; children are no longer fascinating babies and are becoming more active, curious, independent, often troublesome individuals; simultaneously, contemporary economic pressures and inflation make unstable households and casual work arrangements more problematic ways to live (Eiduson, 1978).

The characteristic developmental issue of this age (i.e., the first moves of the child toward independence) may be an issue of particular stress for many of our alternative populations. They themselves in identifying with alternative values have seen their lives as a battle against authority: in some cases, their own parents, in other cases, projected onto society and its values. In fact, their counterculture and antiauthority stance made them of interest to us originally. The idea has occurred to us that the issue of discipline and compliance that emerges during this age period is promoting conflict in our parents and perhaps an unusual ambivalence toward their children. Some describe their children as willful, suggesting, of course, the negativism of the 2's. Parents show a mixture of pride and resentment toward this behavior. There are a number of children who seem to lack the ability to conform to limits during the testing. Whether or not the parental conflict around this developmental period when there is a first show of independence has any relationship to the emergence of hazardous factors remains to be seen.

However, the alternative scene is changing. Alternative parents are becoming reintegrated into the mainstream culture: getting older, moving back from the land, reentering school and the work force. At the same time, mainstream practices continue to embrace some of the alternative perspective: less sexual repression, more interest in sex egalitarianism, a redefinition of family roles, more ambivalence about the roles of parents as authority figures. How will these trends affect tomorrow's child? Alternatives still want their children to be independent and self-reliant early, to be flexible and able to adapt to the rapidly changing conditions of the world, to have control of their own fates, to become competent decision-makers, and to be able to enjoy a full, rich, and well-rounded life. It is hard to quarrel with such goals, but how successfully will they be effected? What impact will family style have on the child? The answers here await longitudinal results.

SOME CLINICAL IMPLICATIONS OF EMERGENT FAMILIES

Children growing up in new families highlight some interesting issues to which clinicians have been sensitized in regard to some American subcultures other than the alternative middle-class child. How does one appropriately assess child behaviors and adult child-rearing behaviors that psychiatrists see as producing deviance or pathology, when these behaviors are valued by the "family" subculture? Or, put more simply, under what conditions does "difference" become "deviance" or "pathology" (Eiduson et al., 1978).

We have learned in recent years from research on minority populations that differences do not invariably connote deviance nor deficit (Deutsch, 1967; Reissman, Cohen, and Pearl, 1964). The clinician's perspective and that of the cultural framework that provides his reference and anchoring points are known to inhibit culture-free inferences and conclusions; this can be true too when dealing with marginal groups, as those under study here. Especially because they have rejected the very values with which clinicians identify, it would be simple—though perhaps unwise and erroneous—to see alternative socialization patterns as "lesser" rather than "differential." The perspective of the cultural anthropologist who tries to identify the competencies and attitudes that serve adaptive and adjustive purposes in another culture or subculture has been helpful in understanding and evaluating differences. Yet, the marginal individuals who comprise emergent families challenge even this enlightened position—for the ways of compromise when one is on the cutting edge of change can be both vagarious and elusive. Study of the young child may turn out to be one of the most effective ways of seeing how adjustment patterns appropriate to different cultural demands, to different ecological levels (e.g., immediate family versus living group versus community), get integrated during the growth process (Alexander, 1976).

Whether what emerges is deviance will have to be looked at from the viewpoint of the service to which it is put in the child's personality—rather than the "foot-in-each-culture" stance from which it was derived. For example, what does it mean for a child to have multiple models for identification rather than nuclear models, when the "family" specifically aims to dilute the child-parental tie and so reduce the possibility of a child's absorption of his parents' neurotic trip? Is confusion in identification, or ambiguity, inevitable—or is family approval so salutary that the child's personality integration becomes enriched rather than fragmented? Are we certain that conflict among adult models for a child is harmful rather than growth-promoting, encouraging opportunity choice rather than ambivalence and manipulation? These are the kinds of issues that changes in values of some contemporary families engender (Eiduson et al., 1978).

The conceptual rethinking that the experiences and values of emergent families demand also raises many clinical issues. The parents, for example, may perceive traditional clinicians ambivalently. As agents of society, clinicians may generate distrust, suspicion, and fear of disclosure; yet persons who are experts in motivation, the unconscious, and personal relationships and who reject obvious social morality are also viewed as compassionate allies. On the ability of the clinician to understand the values and motivations of emergent families will rest their effectiveness in helping these parents and children.

Further, despite their interest in self-understanding, many alternatives are pulled toward environmental manipulation for conflict resolution; thus, their residential mobility, their cycling in and out of the welfare system, and their lack of investment in or commitment to externally accepted goals. It is not completely surprising that a clinician in a child guidance center, after counseling a mother that her child was disturbed by family nudity in their home, reported that by the next session the family had moved to a nudist colony. Some alternatives believe anxieties should not be tolerated; delayed gratification and long-term goals are inappropriate to their "here-and-now" perspective.

Perhaps even more alien to the practitioner's customary orientation is the attitude of some alternatives toward their parental roles and responsibilities. Although some parent as they were parented, comfortable with the conventional, more-or-less authoritative stance, others decry the interventionist aspect of this philosophy. They choose instead to retreat from any intervention or responsibility, feeling instead that a child's fate is sealed by the stars, God, or other inimitable progenitors. Thus, their role is to treat the child as an individual, providing only the necessary nutrients and conditions for growth and making him or her responsible for himself or herself from a very early age (Berger, 1971).

If this way of handling responsibility both on the parents' and the child's part were found only in alternative populations, the implications of this attitude would be limited to a subgroup who might not come into psychiatry at all. However, there is substantive evidence that such stances are beginning to diffuse rapidly from once-counterculture populations into many young contemporary parents in "straight" society (Eiduson *et al.*, 1973). Therefore, some serious thinking about our theoretical positions and our practices seems in order; otherwise we may jeopardize the responsiveness to individual and cultural needs that has always identified our work.

SOCIETAL IMPLICATIONS OF EMERGENT FAMILY STUDIES

Perspectives on society in the 1960s led to the development of the contemporary emergent family styles. They, in turn, as they have voiced their changing values and attitudes in regard to parenting and to child rearing, have found a resonating group in mainstream counterparts. Thus, existing medical, educational, and social institutions have been forced to appreciate that attitudes originally identified as counterculture are already characterizing the parents and children in dominant groups of society whom they customarily serve (Eiduson and Alexander, 1978; Kanter and Zurchin, 1973). The changing attitudes and needs of tomorrow's parents and children which have been identified through this research with young parents indicate a number of target areas for suggested change in existing institutions. To mention only a few of these:

Education

1. Alternative forms of schools and classes, teaching styles, and curriculum.
2. Incorporation of life experiences (outside the classroom) as part of the curriculum.
3. Increased utilization of males as teachers and aides, especially in primary grades.
4. More systematic programming of parent participation in decision making about school policy and programs and in actual classroom involvement.
5. Acceptance of more pluralism in dress, mores, preferences, and interests of children and parents.
6. Deemphasis on grades, performance scores, and IQ tests in favor of individually oriented goals.
7. Development and use of nonsexist books and materials.

Economics and Consumer

1. Establishment of nutritional needs of children, appropriate labeling and advertising; discouragement of availability of nonnutritional packaged foods in public buildings (museums, schools, etc.).
2. Encouragement of simplistic semirural life-styles by making them economically feasible, through developing support systems oriented toward families who try to be self-sustaining and are not productive members of a larger community and through providing training in agriculture, housebuilding, etc., for families who are oriented toward self-help.
3. More flexible financial support arrangements that allow parents who are ambivalent about accepting outside help to cycle in and out of the social agency and social welfare system as their personal conditions necessitate.

Community

1. Giving credence to a variety of religious institutions, in addition to the three major religions, so that children who differ ideologically and by name (and may dress and look different) are not considered deviant.
2. Exposure of adults and children to family style differences so that the two-parent nuclear family is seen as only one of many pluralistic forms.
3. Encouragement of architectural and financial interest in the design and construction of dwellings appropriate to the contemporary extended family.
4. Facilitation of the legal registration of babies born at home and other legal services so that the personal property and rights of parents and children in communal or social-contract arrangements are protected and equal to those in nuclear families.

Many of these issues already have formed a response in contemporary institutions that are struggling to effect change in a world where change, rather than stability, is the order of the day. It is useful to have a group of alternative families to serve as a microcosm for such developments, alerting institutions early to directions in which change is likely to appear, and suggesting the impact on parents, children, and the society at large at such an early point that the needed services and interventions can be thoughtfully considered and planned.

ACKNOWLEDGMENTS

This work is supported in part by the National Institute of Mental Health Research Scientist Career Award No. 5 KO5 MH 70541-05-06 to Bernice T. Eiduson, Ph.D., and by the United States Public Health Service Grant No. 1 ROI MH24947-01-05 and Carnegie Corporation Grant B-3694 and B-3970.

Project co-investigator, Thomas S. Weisner, Ph.D., and senior investigators Jannette W. Alexander, DCSW; Jerome Cohen, Ph.D.,

Irla Lee Zimmerman, Ph.D.; and Max R. Mickey, Ph.D., have also been responsible for directing the project.

REFERENCES

Ainsworth, M. D. S., and Wittig, B. A. Attachment and exploratory behavior of one-year olds in a strange situation. In B. M. Foss (Ed.), *Determinants of Infant Behavior*, Vol. 4. New York: Wiley, 1969.

Alexander, J. Alternative lifestyles: Relationship between new realities and practices. *Clinical Social Work Journal*, 1976, *4*, 289-307.

Ariès, P. *Centuries of Childhood*. New York: Vintage Press, 1965.

Baumrind, D. Harmonious parents and their pre-school children. *Developmental Psychology*, 1971, *4*, 99-102.

Bayley, Nancy. Development of mental abilities. In P. H. Musser (Ed.), *Carmichael's Manual of Child Psychology*, Vol. 1. New York: Wiley, 1970.

Bengston, V. L., and Laufer, R. S. Youth, generations, and social change: *Journal of Social Issues*, 1974, *30*, Parts 1 and 2, 1-163; and 1-205.

Berger, B. Hippie morality, more old than new. *Trans-action*, 1967, *5*, 19-23.

Berger, B. Child-rearing practices of the communal family. Progress report to the National Institute of Mental Health, Bethesda, Md., 1971.

Blois, M. S. Child-rearing attitudes of hippie adults. Progress report to the National Institute of Mental Health, Bethesda, Md., 1971.

Cavan, P. Hippies of the redwood forest. Berkeley, Calif.: Mimeo, 1971.

Coffin, P. The young unmarrieds. In J. S. & J. R. Delora (Eds.), *Intimate Lifestyles: Marriage and Its Alternatives*. Pacific Palisades, Calif.: Goodyear Publishing, 1972.

Cohen., J., and Eiduson, B. T. Changing patterns of child-rearing in alternative lifestyles. In A. Davids (Ed.), *Child Personality and Psychopathology: Current Topics*. New York: Wiley, 1975.

Cooper, D. *The Death of the Family*. New York: Vintage Books, 1970.

Delora, J. S., and Delora, J. R. *Intimate Lifestyles: Marriage and Its Alternatives*. Pacific Palisades, Calif.: Goodyear Publishing, 1972.

Deutsch, M. *The Disadvantaged Child*. New York: Basic Books, 1967.

Diamond, S. *What the Trees Said*. New York: Dell Publishing, 1971.

Eiduson, B. T. Looking at children in emergent family styles. *Children Today*, 1974, *4*, 2-6.

Eiduson, B. T. Child development in alternative family styles: Phase I. Progress report to the Carnegie Corporation of New York: Mimeo, 1975.

Eiduson, B. T. Tomorrow's child. In J. Schwertfeger, (Ch.), *Child Development and Social Welfare*. Ann Arbor, Mich.: University of Michigan, 1978.

Eiduson, B. T., and Alexander, J. The role of children in alternative family styles. In N. Feshbach & S. Feshbach (Eds.), *The Changing Status of Children: Rights, Roles and Responsibilities*. *Journal of Social Issues*, 1978, *34*, 149-167.

Eiduson, B. T., Bernstein, M., and Preston, M. Physical development of 200 children in alternative lifestyles. *Pediatrics*, in press.

Eiduson, B. T., Cohen, J., and Alexander, J. Alternatives in child-rearing in the 1970's. *American Journal of Orthopsychiatry*, 1973. *43*, 720-731.

Eiduson, B. T., and Project Staff. Drug-using parents and their children, Part I; Anticipatory socialization behavior. In report to the National Institute of Drug Use, University of California, Los Angeles, Department of Psychiatry, 1976.

Eiduson, B. T., and Weisner, T. S. Alternative socialization settings for infants and young children. In J. Stevens & M. Mathews (Eds.), *Mother/Child, Father/Child Relationships*. Washington, D.C., 1977.

Eiduson, B. T., Zimmerman, I. L., and Bernstein, M. Single versus multiple parenting: Implications for infancy. Paper presented at meeting of the American Psychological Association, San Francisco, Cal., 1977.

Furstenberg, F. F. *Unplanned Parenthood: The Social Consequences of Teenage Childbearing*. New York: Free Press, 1976.

Gershensen, C. Child development, infant day care, and adolescent parents. *Sharing*, 1972, *8*, 1-10.

Glick, P. C. Some recent changes in American families. Current Population Reports, Special Studies Series P-23, No. 52. Washington, D.C.: U.S. Government Printing Office, 1975.

Gollin, G. L. *Moravians in Two Worlds*. New York: Columbia University Press, 1976.

Greenfield, P. M. What we can learn from cultural variation in child care. Paper presented at the meeting of the American Association for the Advancement of Science, San Francisco, Cal., 1974.

Gross, H. *The Flower People*. New York: William Morrow, 1968.

Judah, J. S. *Hare Krishna and the Counter-Culture*. New York: Wiley, 1974.

Kanter, R. *Commitment and Community: Utopias and Communes in Sociological Perspective*. Cambridge: Harvard University Press, 1972.

Kanter, R., and Zurchin, L. (Eds.). *Journal of Applied Behavioral Sciences, Alternative Institutions*, 1973, *9*, 137-397.

Kaplan, G. *Support Systems and Mental Health*. Baltimore: Grune & Stratton, 1976.

Keniston, K. *The Uncommitted: Alienated Youth in American Society*. New York: Dell Publishing, 1965.

Kinkade, K. *A Walden Two Experiment: The First Five Years of Twin Oaks Community*. New York: William Morrow, 1972.

Klein, Carole. *The Single Parent Experience*. New York: Walker & Co., 1973.

Kornfein, M. *Infancy in Creedal and Non-Creedal Communities*. Los Angeles: Mimeo, 1975.

Kornfein, M., Weisner, T. S., and Martin, J. Women into mothers: Experimental family life styles. In J. R. Chapman & M. J. Gates (Eds.), *Women into Wives, Sage Annual of Women's Policy Studies*, Vol.2. Beverly Hills, Calif.: Sage Publications, Inc., 1977.

Lipetz, M. E., and Davis, K. E. Living together: An alternative to marriage. *Journal of Marriage and the Family*, 1972, *34*, 305-311.

Lyman, M. *Mirror at the End of the Road*. Roxbury, Mass.: American Avatar, 1971.

Macklin, E. D. Heterosexual cohabitation among unmarried college students. *The Family Coordinator*, 1972, *21*, 463-467.

Macklin, E. D. Personal communication, 1977.

Melville, K. *Communes in the Counter-Culture*. New York: William Morrow, 1972.

Miller, P. J., and Sjoberg, G. Urban middle-class lifestyles in transition. *Journal of Applied Behavioral Sciences*, 1973, *9*, 1-294.

Minturn, L., and Lambert, W. *Mothers of Six Cultures*. New York: Wiley, 1964.

Mungo, R. *Total Loss Farm*. New York: Bantam Books, 1970.

Ornstein, R. E. *The Nature of Human Consciousness*. San Francisco, Calif.: W. H. Freeman, 1973.

Otto, H. A. Has monogamy failed? *Saturday Review*, 1970, 62, 23-25.

Rains, P. M. *Becoming an Unwed Mother*. Chicago: Aldine-Atherton, 1971.

Reissman, F., Cohen, J., and Pearl, A. *Mental Health of the Poor*. New York: Free Press, 1964.

Roberts, R. *The New Communes: Coming Together in America*. Englewood Cliffs, N.J.: Prentice-Hall, 1971.

Rogers, C. *Becoming Partners: Marriage and Its Alternatives*. New York: Delacorte Press, 1972.

Roszak, T. *The Making of a Counter-Culture*. New York: Doubleday, 1969.

Sears, R. R., Maccoby, E. E., and Levin, R. *Patterns of Child Rearing*. Evanston, Ill.: Row, Peterson, 1957.

Sears, R. R., Rau, L. F., and Alpert, R. *Identity and Child Rearing*. Stanford, Calif.: Stanford University Press, 1965.

Shorter, E. *The Making of the Modern Family*. New York: Basic Books, 1975.

Singer, M. T. Personal communication, 1978.

Skolnick, A., and Skolnick, J. *Family in Transition: Rethinking Marriage, Sexuality, Childrearing and Family Organization*. Boston: Little, Brown, 1971.

Steiner, S. P., and Maran, M. *Chamisa Road with Paul and Meredith: Doing to Dogs in Taos*. New York: Random House, 1971.

Tannenbaum, A. J. Alienated youth. *Journal of Social Issues*, 1969, 25, 1-5.

Weisner, T., and Martin, J. Learning environments for infants in conventionally married families and communes in California. Paper presented at the meeting of the American Anthropological Association, San Francisco, 1976.

Weisner, T. S., and Project Staff. *Ideology, Values and Family Lifestyles*. Technical report submitted to the National Institute of Mental Health, Applied Research Branch, 1976.

Williams, T. M. *Infant Care*. Washington, D.C.: Consortium on Early Childbearing and Childrearing. Research Utilization and Information Sharing Project, Office of Child Development, 1972.

Wright, H. R. *Eighty Unmarried Mothers Who Kept Their Babies*. Sacramento: State of California; Department of Social Welfare, 1965.

Zablocki, B. *The Joyful Community*. Baltimore: Penguin Books, 1971.

Zimmerman, I. L., and Mickey, M. R. Parental achievement orientation values on Bayley-scored one-year olds. Paper presented at meeting of the Western Psychological Association, Los Angeles, California, 1976.

IV

New Policies for New Families

6

Public Policies and Families

A. Sidney Johnson III

It is important for health practitioners to examine what changing families and the changing ecology of families mean to them and to their work. The thoughtful chapters in this book trace the progression of themes that have been developed: the focus on an historical approach; an examination of social class and its impact on families; a balanced and thoughtful look at not only the needs but also the strengths of black families; a similar look at single-parent families; and a look at newer family forms and policy implications. My chapter fits into the latter part of this sequence. It will focus on public policies as they affect families, both positively and negatively. This focus does not mean or imply that families simply respond helplessly to public policies. Families are agents of change as well. The connection goes both ways.

My chapter will cover two main points: the origin of the family impact concept and the present Administration's commitment to families and family impact.

A. Sidney Johnson III • George Washington University, Washington, D.C.

Let me begin by addressing the question that is the title of the book: *The American Family: Dying or Developing?* I believe it is a mistake to talk about *the* family. We're really talking about families. As pointed out in several other chapters, families are changing, evolving, coping, and doing so with what David Brion Davis termed remarkable resilience and adaptability at the end of his paper. I would also quote briefly from part of Albert McQueen's paper, which he directed primarily to black families and families under stress of poverty but I think can be extended more generally; "Many families cope surprisingly effectively under conditions of racism, poverty, or powerlessness. At the same time, however, all is not well for many others exposed to these same scourges." Families are changing and we need to have a balanced view of what's happening. Not all families are dying, not all are developing. Indeed, many of them are doing a little bit of each at the same time. There is a tremendous resilience and strength, but at the same time, there are great problems, and we need to understand that complexity, although in many ways it appears contradictory.

This whole issue concerning the ways in which families are deeply affected by the environment in which they live—by political, economic, and social conditions—is central to the Family Impact Seminar, which it is my privilege to direct. I believe it will be one focus of the White House Conference on Families as well. This interest in families as they are affected by environmental forces— especially public policies—originated at the hearings that then-Senator Mondale held in 1973, entitled "American Families: Trends and Pressures" (U.S. Government, 1973). The three main findings that emerged from those hearings explain a good deal about the current level of national interest in families and family impact.

The first and obvious finding was that many, many changes are occurring in families, seemingly all at once. These include the dramatic increases in the number of women working outside the home, the trend toward smaller families, the higher divorce rate, the introduction of television into homes, and many more. The point that came through in those hearings is that a whole range of changes are happening at once, and many of us have difficulty dealing with change and ambiguity and uncertainty.

The second main finding of those hearings was that there is a belief in this country that somehow government is neutral to families. That is a myth. We not only have a whole range of public policies, from social security to child welfare to education to day care and the rest, that are explicitly debated and designed to support families. We also have an equally impressive list of policies that are not debated in terms of families but that have direct or indirect, intended or unintended, implicit or explicit effects on families. Many of these effects are mindless and questionable at best. In short, the government is up to its eyeballs in programs that affect families intimately. Consider, for example, the financial incentive in Medicare for institutional care for the aged as opposed to home health care; a welfare law that still permits half the states in this country to withhold federal aid to mothers and children unless the father is absent from home; and the so-called anti-grandmother provision in the day-care deduction, which basically said that you can get a deduction for paying someone to care for your children where both parents or one parent are working, provided that the person who cares for your children is not a relative closer than a cousin. If it's the child's grandmother, it's not a deductable expense; if it's the neighbor's grandmother, it is.

Consider the $750 personal exemption in our income tax, much of which goes for children. Many witnesses at the Mondale hearings said we needed a children's allowance, that the United States is the only Western industrialized country without one. But that is not quite true. We have what one economist called a partial, hidden, upside-down children's allowance in the tax code. It is called the personal exemption. Taxpayers qualify for the same $750 exemption for themselves and their dependents. That sounds very fair and equitable until you realize how exemptions work. The size of the tax savings from an exemption depends on both the dollar amount of the exemption and the tax bracket you're in. So to a wealthy family in the 50% tax bracket, that $750 personal exemption is worth a tax savings of $375. But for a family making only $10,000 that same $750 exemption is worth only about $120-$140 in tax savings. That's a public policy that affects families pretty directly.

Consider the foster-care system, albeit well-intended, that pays more money the further that child is from his natural home. If he is placed in an institution, we'll pay perhaps $20,000 a year or more in public funds. If he's in a community-based foster-family home, we'll pay about $1,500 to $2,000, and if he's in the natural home, we'll pay practically nothing. Clearly this foster-care system provides important help to some troubled families and children, but the economic incentives in it simply have to be reexamined.

Consider the personnel practices of government—its leave policies, transfer policies, the virtual absence of part-time and flexitime opportunities. These are all fundamental policies that affect the ability of a worker to balance his or her responsibilities as a worker and as a family member.

Finally, consider the impact of unemployment on families. If you want to destroy a family in this country, probably the best way to do it is to make sure its wage earner or wage earners cannot hold a job. We take the work ethic in this country very seriously. It has produced tremendous economic gains. It has also produced profound human problems for people who cannot get jobs. The so-called social costs of unemployment are now being studied. Unemployment not only costs financially, in lower tax revenues and increased welfare and related costs, but as health practitioners know better than I, it costs us in human terms as well. A recent study by Harvey Brenner for the Joint Economic Committee (Brenner, 1976) found direct correlations over the last 40 years between increases in unemployment and increases in suicide rates, homicide rates, admissions to mental hospitals, admissions to prisons, and physical diseases related to stress. It does not prove cause and effect, but I doubt that the correlation is just a coincidence. Similarly, when Leonard Woodcock testified to Congress about unemployment, he pointed out that when unemployment rose to 20% in Flint, Michigan, it became the city in our country with the highest rate of alcoholism, child abuse soared, and the drug addiction case loads doubled.

In short, a wide range of public policies—from those in the tax code to those affecting unemployment rates—have direct impacts on families.

The third finding of these Mondale hearings was that the idea of developing what might be called family impact statements—that is, a way to assess and anticipate what public policies like these mean to different kinds of families—should be tested. That is what the Family Impact Seminar—with the help of scholars and policy makers—is doing. We're examining the scientific, administrative, and political feasibility of this idea. Our preliminary findings suggest that it is a very promising but a very complex concept, which makes sense if approached cautiously. And it is amazing to realize how seldom such a fundamental question is raised about public policy. There was, for example, only one issue I can remember in the seven years I worked in the Senate on which some part of the decision was based on an explicit debate of what it meant to families. That was the decision to revoke year-round daylight-saving time. If you will recall, year-round daylight saving had produced the energy savings that had been predicted, but parents around this country would not stand for putting their kids out at a bus stop in the early morning darkness. People can debate whether that decision to revoke was right or wrong, but the key point was that it was debated, perhaps unconsciously, in terms of "What does this policy mean to families?" I think we can apply that lesson to many other policies.

It is, however, a complex issue. There are, for example, a whole range of public policies that affect various families in different ways. There are a whole range of public policies that affect particular members of the *same* family in different ways. Take as an example the question of access for teenagers to sex education, birth control information, or abortion, with or without parental consent.

The last thing we need to do is enact a cliché and apply the whole environmental impact process to families. I believe that it would be a major mistake to create a comparable federal bureaucracy to protect families and to require family impact statements in every government proposal. The whole question of government interest in or intervention into families is rightfully a *terribly* sensitive one in this country. We need to explore this issue very carefully. Our seminar is seeking to do just that by choosing several public policies, setting up task forces, and developing what might be called pilot family impact statements on each. We don't expect to produce definitive impact statements, but we do expect to learn something about the process. In the end of our seminar, we will make recommendations about the feasibility of family impact analysis and develop model legislation, if we decide it would be desirable, to suggest to the local, state and federal governments ways of testing this process.

How does all this relate to health practitioners? I'm certainly not an expert in the field, but obviously Medicare and Medicaid are two health policies with impacts on families. What services to families will they reimburse? Do they contain incentives for home health care versus institutional care or vice versa? Will family therapy, only individual therapy, or no therapy at all be covered in National Health Insurance? I recently learned that, apparently, kidney dialysis is available under Medicaid, at least in some states, only if it's done in hospitals. But the employability of the people who are receiving kidney dialysis while living at home is dramatically higher than of those in the hospital—to say nothing of the cost involved.

And consider all the ways that social and economic conditions of stress that create physical and mental illness affect your ability to help people. You can cure or rehabilitate people, but if you send them back to some of the same situations of stress, over which you have very little control as a practitioner, will they be helped in a permanent way?

These are just some of the issues within the health field that should, in my opinion, be examined in a family impact framework.

Now some people wonder if this family impact approach is anti-individual. Does it state, they wonder, that the needs of children, or adolescents, or women, or men are less important than the needs of the families? The answer is no, categorically. What the family impact perspective is trying to say is that all individuals live within networks and that one of the most influential is typically the family. Thus, if we want to effectively deliver services to those individuals, we have to be aware of the ways in which they're impacted by the networks they live in, their families, and the ways in which their families are impacted by public policies.

The White House Conference on Families, to be held in December of 1979, is expected to look in part at the ways in which public policies affect families. Hopefully it will also look at the ways in which other major institutional forces affect families, including the media, the world of work, the court system, private organizations, and others. This book can be very helpful in encouraging a reexamination of the ways public and private policies affect families. But we need to have realistic expectations about what White House conferences can do and what they can't do. They cannot enact legislation. They cannot appropriate funds. They cannot be long-term lobbies. They are not that kind of mechanism. They can, however, stimulate interest. They can lead to increased resources. They can change attitudes, but they're not a one-shot solution, and we need to be mindful of that fact. All of us in our work need to be aware, as this book underscores, of the ways in which families change; the strengths as well as the weaknesses that families experience. That is the kind of balanced perspective that is being increasingly reorganized by individuals and organizations concerned about families and that will, I believe, be reflected in the White House Conference on Families.

I'm very hopeful that this kind of emphasis, augmented by books like this one, is going to lead us to a point where the services and opportunities we seek to make available to individuals will be more effective because they are built on a true appreciation of the diversity and integrity of family life in this country.

REFERENCES

Brenner, Harvey. *Estimating the Social Costs of National Economic Policy: Implications for Mental and Physical Health, and Criminal Aggression*. Washington, D.C.: U.S. Government Printing Office, October 1976.

U.S. Government Printing Office. Hearings before the Subcommittee on Children and Youth, September 24, 25, 26, 1973.

7

Actions and Services for the New Family

Marvin B. Sussman

INTRODUCTION

In the 1970s, as in past decades, the same question is being raised: Will the family survive as a primary social system in the 21st century? The answer is a qualified yes—in that some form of family will persist, but the meaning it will have for its members, the structure of relationships, the way in which marital and parenting functions will be performed, and its durability over time are unknowns. Also, these future functions will vary extensively from the more commonly known and prevalent nuclear family form of procreation, where males traditionally hold gainfully employed roles and women are homemakers.

Mass-media popularization of unusual life-styles, such as cohabitation, and experimental family forms, such as group mar-

Marvin B. Sussman • Bowman Gray School of Medicine, Winston-Salem, North Carolina.

riage, has tended to sharpen our awareness of less esoteric variations that are "spinning off" from the traditional family and have potential drastic consequences for the life-style of its members and the organizations and institutions of society. Popularization has salutary effects, however. First, in examining novel phenomena, behavioral scientists become sensitized to the range of family forms and associated behaviors along an experimental-traditional continuum and are better able to identify events producing or forcing these changes. Variations become more obvious; thus, our conceptualizations and analytic schemes come closer to reality. Second, deviation from the mode is a message of import in itself. It suggests that things are not working out for the traditional family and that the situation is sufficiently fluid so that the family must examine its purpose, meaning, and activities and make decisions on how to use its limited energies and resources.

The phenomenon of family pluralism is more apparent now for several reasons. The frequency of forms such as single-parent and dual-work families rather than the traditional nuclear or single breadwinner has increased over the past several decades. Some of these so-called variant forms are becoming more visible as behavioral scientists and demographers research these developments. Also, families are beginning to organize and make highly vocal demands. For example, springing up across the nation are new organizations of working mothers, welfare mothers, retired couples, and single parents that have transformed once-isolated individuals and families into social groups offering support to each other and exerting pressure on government agencies and human service institutions for needed services and removal of discriminatory government policies and programs. Academics, professionals, governmental officials, and the general public are confused about the implications of these alternate forms and are sometimes bewildered about taking action either to support or to suppress them.

ISSUES AND PROBLEMS

Although knowledge of the consequences of multiple family forms is scant, the general position is that in the 1970s pluralism in family structures reflects variations in the racial, ethnic, religious, and age groups that compose the American salad-bowl culture. Each of these forms, because they are so different from one another, has varied problems to solve and issues to consider when dealing with organizations and institutions and its own internal family relationships. Most important, variant family forms have different needs from the traditional nuclear family for outside services and supports, which for the most part remain unmet in the United States today.

Until recently, we have used an idealized family form—the single-breadwinner, intact family—on which to base public policies and programs, ignoring or deprecating other family forms that differ from this "ideal" type. An example is the Chicano family who have an historically strong extended-kin network. Their representatives present the very articulate case that in addition to their being denied services, some new government programs—that actually are trying to help—are, in fact, destroying their family system. Realizations such as these have encouraged the investigation of new ways to create organizations, institutions, and legislation around people rather than to fit people into the ongoing societal system, where the traditional nuclear family of the single breadwinner is considered if not the sole, at least the ideal and healthy form. Much too often, bureaucratized service systems function largely for the benefit and "efficiency" of the staff and neglect either to offer services to the wide array of family types or to vary programs and procedures so that relevant and needed services are geared to the special needs of each type. This is indeed a difficult problem to remedy once bureaucratic practices are fixed. It is much easier to continue to act as if families share a uniform structure and then to proceed to offer a unified program. Individual family types suffer in the process.

DEFINITION OF TERMS

It seems appropriate to describe briefly what are variant family forms (see Table 1).

One can view family structure from another perspective, household composition. James Ramey utilizes the typology seen in Table 2.

TABLE 1

FAMILY TYPE BY ESTIMATED PERCENTAGE OF DISTRIBUTION IN THE UNITED STATES, 1976[a]

Family type	Estimated Percentage	Distribution
1. Nuclear family—husband, wife, and offspring living in a common household ("intact")	45	
a. Single career		30
b. Dual career		~5
(1) Wife's career continuous		no estimate
(2) Wife's career interrupted		no estimate
2. Nuclear dyad—husband and wife alone, childless or no children living at home	15	
a. Single career		4
b. Dual career		11
(1) Wife's career continuous		no estimate
(2) Wife's career interrupted		no estimate
3. Nuclear family—husband, wife, and offspring living in a common household (*remarried*), no estimate of career patterns	10	
4. Single-parent family—one head, as a consequence of divorce, abandonment, or separation (with financial aid rarely coming from the second parent), and usually including preschool and/or school-aged children	15	
a. Career		11
b. Noncareer		4

5. Kin network—three-generation households or extended families where members live in close geographical proximity, operating within a reciprocal system of exchange of goods and services.   2

6. Other single, widowed, separated, or divorced adults   7

Emerging experimental forms
1. Commune family   6
   a. Household of more than one monogamous couple with children, sharing common facilities, resources, and experiences, socialization of the child is a group activity.
   b. Household of adults and offspring—a "group marriage" known as one family—where all individuals are "married" to each other and all are "parents" to the children. Usually develops a status system, with leaders believed to have charisma.
2. Unmarried parent and child—usually mother and child, where marriage is not desired or possible.
3. Unmarried couple and child—usually a common-law type of marriage, with the child their biological issue or informally adopted.

100

[a] Reformulated from Sussman, 1974a. Based on communication with James Ramey, who suggested inclusion of Item 6 (singles), based on the 1970 Census data on all adults.

Table 2
ADULT HOUSEHOLD COMPOSITION BY ESTIMATED PERCENTAGE OF
DISTRIBUTION IN UNITED STATES, 1976[a]

Type		Percent
1.	Adults living in single-breadwinner nuclear families	13
2.	Adults living in dual-breadwinner families	30
3.	Adults heading single-parent families	17
4.	Adults living in extended families	4
5.	Adults living in expanded families	4
6.	Other single, widowed, separated or divorced adults	21
7.	Adults living in childfree or post-child-rearing marriages	11
		100

[a]J. Ramey *Marriage and Family Review*, 1978, 1 (1). Bureau of Labor Statistics. *News Release*. USDL 77-191, March 8, 1977.

At first glance, there appear to be discrepancies between Tables 1 and 2. The difference can be explained by the availability of more recent 1975 Census data and differences in the categories of the typologies.

In the past three years, there has been a dramatic increase in the number of dual-work families: today a little over 65% of wives with husbands present and children in the home are working wives. Single-parent families are growing at 19 times the rate of dual-parent families, with a 73% increase in the number of women heading single households since 1970. And a continuous drop in the marriage rate and rise in the divorce rate and cohabitation accounts for the 21% of adult households where the individuals are single, widowed, separated, or divorced. The major discrepancy—single breadwinner, 30% (Sussman) and 13% (Ramey)—may be explained by use of the "Adults living in childfree or post-child rearing marriages" category by Ramey and not by Sussman. An unknown percentage of these may be single-bread-winner families.

VARIANT FAMILY FORMS

Families vary in their life-styles, ambitions, marital health, social and intellectual competence, physical health, conflict and its resolutions, decision making, and values. The list could continue on and on. The typology presented focuses on the positions or roles present in each family type, suggesting links among these positions. The typology presents the players rather than the drama itself. The presence or absence of positions that comprise each form, however, has important ramifications for family dynamics, need for human services from organizations and social networks, mass media and education programs, and marital and family health.

Individuals move continuously in and out of these family forms. Expected is an increase in the incidence of dual-work families with a leveling off of single-parent structures at between 16% and 18% by 1980. Dual-work and single-parent families will make up the major nontraditional forms. The reasons for these trends include the increased needs and opportunities for women in the job market; service organizations that can assist families where a single parent or both partners are gainfully employed; the continuous trend toward removing discriminatory and denigrating laws covering separation and divorce; high rates of divorce by couples with children; changing values; a growing population of women aged 15-44; and improved medical and household technologies. While the leveling off of single-parent families is expected at about 20%, it is a family type, largely female-headed, beset with innumerable psychological, social, and economic problems.

CONDITIONS PRODUCING VARIANT AND EXPERIMENTAL FAMILY STRUCTURES

Numerous explanations of the increasing incidence of variant and experimental family forms, or at least their visibility, are being offered. One explanation emphasizes that institutions that still support the traditional nuclear family are reaching the limits of their capabilities to prevent the rise of these new forms. There are too many individuals who are opting out of traditional marriages. One political solution is to recognize pluralism by increasing the options for individuals and families of different forms. The defunct 1979 White House Conference on Families recognized family variation. When the plans for the White House Conference on Families were announced by President Carter in an official statement January 30, 1978, the President stated that "the main purpose will be to examine the strengths of American families, the difficulties they face, and the ways in which family life is affected by public policy." This recognition of multiple family forms can be viewed as one form of intelligent cooptation.

A second explanation is based on the theory that U.S. society and other Western ones like the German Federal Republic, Great Britain, Sweden, and Denmark are in the throes of a social revolution. The current attacks, especially by the young, on old-order institutions such as the traditional family are being expressed through revolutionary experiments in communal living and group marriages. Catching on with some youth of the new generation is the pattern of declaring the nuclear family obsolete and discarding this form by seeking an alternate life-style—one counter to their parents' with its supposed boredom, tastelessness, meaninglessness and even agony. Some form of communal living is what these young people want. Based on a new value system, these and other "way-out" forms represent opposition *to* and confrontation *with* the old culture's way of life. The values of living with one's peers, emphasizing the present, and experiencing life as individual fulfillment have replaced values of generational responsibility, future-orientedness, and individual striving.

A third explanation is macrosocietal. New family forms are viewed as emergent from the ever-changing pattern of linkage between families and their members with societal organizations and institutions. Societal structures, values, and activities have often created conditions that have fostered variant forms. Over the last three decades, the press for women to enter the labor market has increased the prevalence of dual-work families. For example, the incidence of single-parent families, as demonstrated, has risen and is related to changing perspectives regarding divorce, rising illegitimacy among the never-married, and continuing high rates of divorce by couples with children. In the case of the poor, discriminatory legislation, policies, and practices of welfare systems such as Aid to Dependent Children have contributed to this rising incidence. *However*, societal institutions have not kept pace in their activities with such changes in family forms when one reviews their current policies and programs.

BASIC VALUE CHANGES

Basic values and changes are occurring within the society as well as within the family. The right to separate and file for divorce in order to dissolve a marriage, a marriage that is creating an earthly hell for both partners, is becoming an accepted new value. About 28 states have some version of a no-fault divorce law, which eliminates divorce as an adversary process, where one individual must be found guilty and the other partner innocent (Glick, 1975). Another critical value change is that human sexuality has been removed from the back room, and there is open discussion and improved education and training, so that each member of the society can now discover himself or herself as a human sexual being.

Living together without being married officially, called *cohabitation* by some family watchers, is increasing in incidence. Unrelated adults sharing two-person households increased from 242,000 as reported in the 1960 Census to 327,000 in 1970. A 1970 Census report indicated that 1.3 million persons were living in two-person households. The more significant point is that such arrangements are being "legally sanctioned." Consider the Lee Marvin case, where the cohabitee of the famous actor was granted by the California court a share of the couples' equity upon separation of the partners. This is an example of a proposed legal shift: cohabitation is rivaling marriage.

The sharing of responsibilities and parity in dyadic relationships within the family is a critical new value. This is evidenced by the emergence of the new phenomenon of the personal marriage contract. The contract sets forth the marital expectations of each

partner in the marriage relationship. While equality in all relationships is empirically impossible and may not even be desirable, the
personal contract attempts to undo what has been historically
found in the American family, namely, a male-dominated marriage
system. Taking off from the marriage creed of "for better or for
worse," it attempts to share equally what is good and what is bad
in a marriage partnership or a cohabiting relationship. The contract
may turn out to be the instrument to save dyadic marriage
(Sussman, 1974b). Within the context of a society that supports
partnership agreements or contracts, and one that has accepted the
norm of zero population growth as indicated through increasingly
lower fertility levels, there is, on the psychological level, an increasing sense of individuation. The individual, rather than the family,
is being emphasized as the primary unit.

Concomitant with the emphasis on the individual is an increasing
pervasive feeling among an increasing number of the population
that parents as well as children have a "right" to live a full and
high-quality life. This is largely because of an alteration of psych-
osocial geometries—a shift in time and space perspectives. In regard to time, there is an obvious shift from viewing life in a generational continuity perspective to one of viewing the totality of life
within one's own life span. The shift is from viewing one's contribution to familial and cultural history contributions to the present, today not tomorrow. This is accomplished by picturing one's
own life, family, and friends not as phases in a continuing process
but properly judged in isolation for their intrinsic quality. It is in
this changed time perspective that social movements, including the
human rights and gender-role movements, have their ideological
roots. It is in today, not in yesteryear or tomorrow, that one
searches for the good life. The individual is to experience the quality of life and to obtain a right to such experience. The corollary is a
reduction of faith in traditional family forms and institutions along
with increased suspicion of authority and reduced faith in prom-
ises and the future. This attitude is shared widely in our society
and influences the behavior of family members.

The spatial shift involves emphasis on collegial links among
peers, with lessened reliance upon the superordinate-subordinate
model of relationships as found in traditional organizations, in-

stitutions, and primary groups such as families. Individuals desire lateral over vertical relationships and make an effort to reduce the power and authority of superordinates in vertically structured organizations, including traditional families. Such techniques and processes as noncompliance, nonparticipation, influence, and strike are used in order to restructure existing role relationships.

Another spatial shift is the turning away and being turned off by the world of superstructures, institutions, and organizations and historic and traditional ideologies and values, with the movement toward self, a shift from outer to inner space. The search is for an answer to the question, "Who am I?", the relevancy of immediate experience, the need to become a quality person through experimentation in a not-very-high-quality world. There is an implicit assumption that there are optional ways of living, and one finds the one appropriate for oneself. Traditional organizations and ideologies may work for some but not for all. One has the option to choose and pursue at one's own pace a life-style and a family form that from the individual's viewpoint will bring about the greatest self-actualization.

TRENDS

1. The divorce rate will continue to rise, with a corresponding drop in the marriage rate. Between one-third and one-half of all marriages in 1977 are likely to end in divorce, the majority occurring during the first three years of marriage.
2. A continuous rise in singlehood is expected. This is a consequence of increasing decisions not to marry; increased rates of cohabitation—basically a nonlegal marriage form; and a longer postponement of remarriage among the divorced.
3. There will be a continuous increase in the incidence of childfree marriage, with major efforts to sanction this union as a socially approved family form. The costs of having and supporting children; increasing emphasis on equitous dyadic relationships and the importance of the marital health of couples; and the increasing incidence of sterilization—which, incidentally, is the preferred form of contraception in the United

States and worldwide—are reasons for the rising number of childless marriages.

4. Political and legislative efforts will be made to "stabilize" the family of the future. The proposed 1981 White House Conference on Families will be one such beginning effort. The actions to date are too preliminary to forecast the ideological cast of the conference. Will family pluralism be used as its conceptual base, or will there be efforts to reestablish a nuclear family with a traditional role structure?

5. Increasingly the "everyday" family or local family (Ramey, 1977) will become the significant primary group for the individual, perhaps even eclipsing the legal family in psychological and social importance. This is the group of individuals who may or may not be related by blood or marriage and who provide intimacy, emotional support, companionship, role models for children, and help when in need. These are the persons one can count on, to whom one has free and easy access, and with whom one likes to spend free time.

POLICY RECOMMENDATIONS

Introduction

Given these trends, there are a number of policy recommendations that I hope can be given serious consideration. They are of two types. One is concerned with handling the issues and problems endemic to the conditions of variant family forms and marriage styles. Recommendations are therefore to ameliorate such conditions. A second type is concerned more with prevention. It proposes more basic and generic change in programs and practices.

Role Models for Children. There is a plethora of studies on paternal deprivation and its consequences for the psychological, social, and physical development of children. While many of these studies are suspect in relation to selection of subjects and matched controls, nevertheless there appear to be significant correlations between high levels of deprivation in youngsters in relation to their social, psychological, and educational development and the absence of male role models, whether in single-parent or intact families (LeCorgne and Laosa, 1976; Jantz & Sciara, 1975). Experience with another adult, especially a successful one, enables the transmission of a body of experience and folk knowledge that facilitates the child's development of interpersonal competence, skills in coping and even manipulating the normative demands of bureaucratic organizations, and realistic and achievable goals. Organizations such as Parents without Partners have come into being as a consequence of the single-parent phenomenon. Their programs are largely oriented toward adults and are social in nature.

Required now is a public policy expressed on the community level that would permit some reallocation of public and voluntary sector resources that will enable *children who need adults to have partners without parents*. The aim is to develop a pool of eligible adults who could function as members of an "everyday family." At first blush, this recommendation may seem minuscule as a thunderous public policy. Upon earnest reflection, however, it calls for a massive reorientation of how we define and perceive family. It requires, furthermore, an opening up of traditional, albeit legal thinking and definitions.

Divorce Insurance. The lack of adequate income is the primary problem found in the overwhelming majority of female-headed households. For many years, a variety of income maintenance proposals have been suggested for families where the members are poor. To create an effective income maintenance program entails a major social reform, with probable crisis and conflict between those who stand to benefit and those who will have to pay for such programs. Undoubtedly the costs of such a major program would be substantially greater than the cost of current programs to alleviate poverty. Alleviation is the outcome of income maintenance programs rather than the eradication of poverty. Guaranteed-annual-income plans and negative-income-tax schemes are important for that segment of the population who would take advantage of opportunities to be counseled and trained for jobs and to use the variety of social and rehabilitative services available in the society. There are too few programs throughout the United States that integrate these various components, and it is very unlikely that if we do achieve an income maintenance program—and I doubt that we will—it will be packaged as I have briefly described.

We have this fact, and an ancient but important finding reported in 1977, upon which to base a traditionectomized policy recommendation. The finding is that "Earning one's livelihood for oneself and children is beautiful." Stated more scientifically, the divorced, separated, or liberated mother who assumes control over her own destiny—that is, personal control—and proceeds to earn her family support is more likely to be better off than the mother who stays at home and depends upon unreliable, unstable, controlling, or stigmatizing sources of income (Bould, 1977). Income that has a clear legal, moral, and practical source is probably very important in providing mothers the freedom and autonomy to plan for their own lives, as well as for the lives of their children, with some sense of assurance and certainty. On the basis that a welfare dollar is not the same as a work dollar and that child support payments are not the same as social security payments, I am advocating a public policy that involves largely the private sector of our economy. I am advocating *a policy and program of divorce insurance*.

I have tested this notion of divorce insurance with a number of people in the industry and while they think it is a bit far-out, they at least have not considered me crazy. Because of the pressure of other commitments and establishing other priorities, I have not worked with actuaries regarding the statistical/actuarial bases of such a program. However, I think that they are workable. In my plan, I adapt the family income insurance principle. A family income policy requires a relatively high premium during the early years of the policy, so that in the event of the premature death of the parent, the surviving children will have a substantial income. After 20 years, the policy turns to ordinary life insurance: payment upon death is at the policy's face value, and the holder pays a lower premium after 20 years. In divorce insurance, the premiums would be very high at the commencement of marriage, with payments minimal if divorce occurs during the first two or three years of marriage. As the marriage continues, the insurance payments would decrease and the benefits would increase. The principle is to protect the single parent who has major responsibilities for child rearing as a consequence of divorce; to prevent this person from being ripped off; and to provide substantial resources so that this individual can be retrained for work, if this is her desire, or have a substantial amount of money to carry her—guaranteed funds—until she remarries. This could be a jointly sponsored program between the private and government sectors; it can be part and parcel of fringe benefits worked out through a union program or a company-controlled one. A computerized well-monitored system to exclude rip-off artists who simply marry and then get divorced in order to collect insurance is required. While the premium structure should diminish some of this activity, the use of computers and developing a list of bad risks, just as Master Charge does, should handle this overwhelming problem.

Insurance income, like Caesar's wife, is above reproach. Such income provides the resources, as I mentioned previously, for women to go into the open market to obtain the necessary training, better jobs, and adequate day-care services. If the private sector is unable or unwilling to take this program because of its high risk, and nobody wants to function like Lloyds of London, then such a no-fault divorce insurance could be organized similar to the way we handle Social Security, through the payroll tax. In fact, there may be more interest among those in the work force in such an insurance program because it directly affects them. By this, I mean that most people who are in the productive years of their lives are the most prone to divorce. In contrast, those of us who are now contributing to the Social Security system through the payroll tax are doing so with the hope that when we retire the generations that succeed us in the work force will be magnanimous and pay us our pensions. Let me assure you, as economists have assured me, that there is no guarantee that what we put into the Social Security system is there for us to use. What we will get upon retirement, if anything—and I hope that we do—will depend upon those who succeed us in the work force. No-fault divorce insurance needs your serious attention.

Utilizing Extended and "Everyday" Family Members as Caretakers and Socializers. Adequate day care has been given a very high priority by human service professionals who are concerned with working-parent and single-parent households. Efforts to get an adequate day-care bill passed since the Nixon presidency have been unsuccessful. David Robinson (1977) in a recent report indicates that day-care centers, funded or operated by private industry, have obtained only a very limited acceptance in the United States. Most parents who work prefer to have a relative or another mother look after their children while they are at work. Costs may be a factor. It is estimated that home-care arrangements cost approximately $800 a year nationally, while formal day-care centers average $2,000-$3,000 per year. Currently federal expenditures are about 1.5 billion dollars for day-care activities. Approximately one-third is for noncenter day care, and the remaining funds are subsidies to support teacher training, loans, meals, day-care centers, and Head Start.

One breakthrough in this area was the federal income tax credit for child-care costs enacted in September 1976. It allows a tax credit of 20% of employment-related child costs up to $2,000 for one child or up to $4,000 for two or more children, with no income ceiling. I have spoken to only one portion of the problem and have not at all looked at the quality of such environments. Let me assure you that they vary extensively as reported in the literature. While it is too early to evaluate the impact of the federal income tax credit upon the utilization of day care, let alone the quality of day care, it is nevertheless important to establish whether such a public program is assisting parents with their day-care needs. This discussion has been by way of introduction to the recommendation.

My inclinations are to provide *an option whereby the child of the working parent may be able to live in a creative daytime environment provided by a relative, a member of that child's extended or everyday family.* Public policy would provide a program with adequate funds for the training of adult relatives for socialization roles—largely to make up any deficiencies that may be evidenced by appropriate testing—and for packaging such a program with appropriate economic and social supports. In addition to providing income to such caretaking families, the program would work in coordination with existing day-care centers so that such children can, on a weekly basis, go to such day-care centers in order to obtain a different kind of group experience. Such formal day-care centers could provide backstopping services to handle the child normally taken care of by the relatives, if and when the relative is unable to do so because of illness or vacation or for some other important reason.

There are operational models for such a program. In our own

research during the past four years, we have been exploring the feasibility of providing such economic and social supports to member units of extended families in order for them to provide a creative environment for elderly and chronically ill members.

Taking a child out of the home and providing for his or her care is only one possibility and may not be the best one. What are the prospects of taking care of children of working parents on their own turf? This may be too prohibitive in cost but may be feasible if the family is reasonably large and an individual could be trained as a "homemaker-health-care worker" to do homemaking chores and other supplemental services (Shinn, 1976). It may be economically feasible via the tax credit mechanism, by direct subsidy from the government sector, or as part of the program of a human service agency in the voluntary sector. An example I have used is taken from the birthquake of programs being developed for the care and maintenance of the elderly in our country. *My policy recommendation is that such a program applied to the elderly and the disabled could be modeled for varied family forms where children need care.* I would further suggest in such a policy that attention be given to the utilization of retired persons, those who are bright in mind and body, who could provide a lifetime of experience and well-honed skills not only in housekeeping and maintenance roles but as educators, intimates, and friends of children.

Prevention

Recommending public policies aimed at prevention causes me a good deal of worry. A policy, as you will recall, is essentially a static document. It states what ought to be and functions as a constraint. It runs counter to my notion that there are multiple pathways to heaven, or at least to the good life, and that people should have a choice of these options. My view of a "modern" society is one that presents more and more options for more and more individuals. Consequently the following public policies and programs are presented with great hesitance and some self-doubt and skepticism.

Universal Training for Competence to Deal with Bureaucracies. A colleague has defined *bureaucracy* as "turning creative energy into solid waste." Whatever may be your definition, we are faced daily with dealing with bureaucracies. Since one-third of all marriages will end in divorce, the majority by the time they are three years old, one has to deal with a number of bureaucracies and institutional systems as one exits from a marriage. Consequently, there is a great need for universal training on how to deal with and even manipulate bureaucracies in order to survive. This is more a program than a policy, but it may become a policy if it is mandated as part of everyone's education just as one is required to have minimal skills in reading and writing in order to graduate from public school.

Qualifying Test for Marriage. Child-free marriages are increasing in incidence. We estimate in 1977 that 23% of all households are child-free because individuals have chosen not to have children or are in the later life-cycle stage known as the "empty nest." While the incidence has been increasing in recent times, the incidence of child abuse, if not increasing, is at least becoming highly visible. It does suggest that there are many people who should not have married in the first place and, if they did get married, were not fit to have children. Consequently, I would support a *public policy that would require passing a qualifying test in order to obtain a marriage license.* Included in the preparation package would be a personal marriage contract with basic provisions regarding economic, work, and child support functions—provisions that are likely to be held valid by the courts—along with the option for expert medical and personal counseling regarding child spacing, birth control, and parenting. As Clark Vincent has pointed out repeatedly, getting married and having children is considered a "natural" phenomenon. Yet we take instructions to learn how to drive a car, and a license is required. As we get older, we are subject to repeated examinations to establish our abilities to drive a car in order not to injure ourselves or others. We are constantly signing contracts and other documents to demonstrate our competence, whether it is to get a passport and a visa to visit another country; to enroll in a public school by establishing that we are healthy and noncontagious; or to qualify to have an ID card in order to buy a beer or visit a singles bar. Such licensing, contracting, and educating as I propose will act as a deterrent for many individuals and prevent them from getting into a marriage where they may create a holy hell for one another. The process may enable others to work out the differences between them, establish common expectations of what a marital and family relationship will and ought to be, and provide the less able member—less able in ability to obtain gainful employment—with adequate resources so as not to become a public charge and a mental case.

The evidence that I have accumulated on marriage contracts suggests that marriage has been taken seriously from almost the time of recorded history. Licensing, not in the modern sense, was effected by the family, the kin group, or the moiety. These provided the institutionalization and sanctity of the marriage instead of the state. Such a public policy may appear to be constraining, and upon reflection, I believe that it does act as a constraint or deterrent. Looking at it optimistically, I believe that it provides the option of getting married and being serious about it or not getting married at all. These seem to be reasonable alternatives.

It is unlikely that the society is ready for the recommendation I just made. Consequently, I have to fall back on a safer recommendation. Recognizing that operant conditioning is acceptable in some quarters and has demonstrated some limited success, the following recommendation is a rather general one. Beginning from the day the child leaves the womb, we need *educational programs built along the lines of "directed socialization,"* programs that will not only modify sexist attitudes but change women's feelings and attitudes about themselves. Such programs would point to the fact that women who marry are likely to experience divorce and that there will be a need for them to be self-providers. Such educational programs would consider not only the practice of contraception but perhaps the postponement of marriage. Contract marriage would be an important part of such an educational program. If marriage occurs and children are born, the notion of joint parenting would be a normal expectation. With divorce being inevitable, innovative demonstrations are required regarding custody. Is it possible to develop new custody programs involving both the male and the female partner? Perhaps if non-rip-off economic settlements upon divorce can be established, the current endemic hostility, anger, and alienation can be mollified sufficiently to foster the parity custody of children.

I am somewhat fearful that such recommendations for change

are difficult to come by in the course of normal development in a society that functions on principles of evolution rather than revolution. Also, I stand the danger of being labeled as a nontraditionalist for my advocacy of pluralistic family forms and concomitant policies. Whether you provide me with strokes or pricklies, I find that it is better to be a lion for a day than a mouse for a lifetime.

REFERENCES

Bould, S. Female-headed families: Personal fate control and the provider role. *Journal of Marriage and the Family*, 1977 (May), 339-350.

Glick, P. A demographer looks at American families. *Journal of Marriage and the Family*, 1975 (February), 15-26.

Jantz, R., and Sciara, F. Does living with a female head-of-household affect the arithmetic achievement of black 4th grade pupils? *Psychology in the Schools*, 1975, 12 (October), 468-472.

LeCorgne, L., and Laosa, L. Father absence in low-income Mexican-American families: Children's social adjustment and conceptual differentiation of sex role attributes. *Developmental Psychology*, 1976, 12 (September), 470-471.

Ramey, J. Sexual behavior and alternative lifestyles. *Society*, 1977, 14 (3).

Robinson, D. Working parents choose home-based child care arrangements. *World Work Report*, 1977 (February), 7-12.

Ross, H., and Sawhill, I. Time of Transition—The Growth of Families Headed by Women. Washington, D.C.: The Urban Institute, 1975.

Scott, R. A. The selection of clients by social welfare agencies: The case of the blind. *Social Problems*, 1967, 14, 248-257.

Shinn, E. Report on Phase I of the case management study. New York: National Council for Homemaker-Home Health Aide Services, Inc., 1976.

Sussman, M. B., and Associates. Report of Forum 14, Changing families in a changing society. Washington, D.C.: Government Printing Office Report to the President, White House Conference on Children, pp. 225-238, 1971.

Sussman, M. B. Family sociology: Issues and developments in family sociology in the 1970's. In Margaret S. Archer (Ed.), *Current Research in Sociology* The Hague: Mouton, 1974. (a)

Sussman, M. B. The personal contract study, in progress. Supported by a grant from the Russell Sage Foundation, 1974. (b)

U.S. Bureau of the Census. Marital status and living arrangements: Current Population Reports Series P-20, no. 306. Washington, D.C.: U.S. Government Printing Office, 1977.

7

Practical Considerations

T. Berry Brazelton

I found Dr. Sussman's paper absolutely fascinating. The most fascinating aspect was not the statistics (which were extremely interesting in their own right) but the attitude he had toward them. Indeed, we have been sizing up the changing family against an old expectation. All of our bureaucracy has been aimed at this expectation, and our consideration of changing families was being measured against an expectation that wasn't going to work.

For a long time, our society has been sticking a Band-Aid on whatever oozes, not with any overall goal but just out of a kind of cultural guilt and denial. My guess is that we had enough going that looked pretty good, and we could thereby foster our denial. Our societal efforts have been grossly inadeqate to the underlying needs, and this failure becomes a reinforcer for failure. I'm very concerned about any actions we might take to support families—single-parent families, new family forms, any kind of family—that don't make personal responsibility, ultimate responsibility, the

T. Berry Brazelton • Harvard Medical School, Cambridge, Massachusetts.

major goal. Consider what is now being done—Aid for Dependent Children, foodstamps, welfare, and so on. I almost challenge you to name anything that we now do for families that isn't a reinforcer for failure. Each of these programs is couched in the implicit message that "Since you've failed, you deserve our attention."

With that in mind, I find it absolutely critical that we wonder what we are doing when we reinforce the failure. Not only are we saying you really can't make it without society's largess, we're also saying that we know the answers and you don't, so stay down there with your hands out and then we're going to be able to help you.

I wanted to be a missionary until I was about 12 years old— until I met one. She had just come back from Africa and she said, "Oh, I'm helping those poor Africans." That was the last word I listened to, as I changed my mind about being a missionary. But I think we are all trained to be, in some sense, missionaries and "hander-outers," and we are in the process of reinforcing people for their inadequacies and for their failures at a time when we can no longer afford to do this.

Let us consider some possible new avenues of activity that would provide support for the family. The women's movement, for example, is one of the most exciting, important, and necessary things that has happened in our generation. But for the women's movement to ignore women's nurturing role and just concentrate on their more career-oriented roles is perhaps, again, applying just one more Band-Aid. In this case, it is a Band-Aid aimed at upgrading women's opportunities for equal pay and for equal chances at careers and is an attempt to negate the built-in advantages for males in our career-oriented society. Such unequal chauvinistic practices must indeed be changed. But to emphasize this one role for women at the expense of the other—of nurturing a family— may be more destructive than it is constructive. Why can't we emphasize and reinforce women toward two roles? As soon as we begin to reinforce the women's movement and support the claim that an adequate person can deal with two roles successfully, I think we will have strengthened the whole structure of the movement.

Men's capacities to deal with two roles successfully needs to be reinforced as well. We need to begin to support men so that they will believe that they can have a successful career and still perform in the nurturing role of father; they *can* be good at both. I believe the sharing of nurturing roles is absolutely thrilling. Many men, however, are so stereotyped in their perceptions of role sharing that to be seen as the nurturer is very disturbing to them.

I recently saw a young man with his 2-year-old daughter, for whom he has been the primary caretaker. I asked, "How's it going?"

He said, very unenthusiastically, "OK."

I said, "OK? That's not so good. What do you mean OK?"

And he said, "Well, she's beginning to talk, and she's got me all confused. She calls me Mama and Daddy. She calls me Mama when she needs something and Daddy when she wants to play."

I said, "Oh, that's wonderful. She really knows how to get what she wants from you."

"That's OK", he responded, "except when my male friends are around."

Now, isn't it tragic that this man, who obviously has a good relationship with his child, can't really tolerate this kind of role crossover and the self-questioning it engenders in him?

Another potential Band-Aid in our society, one that really concerns me, is the direction that day care for small children appears to be heading. Day-care opportunities can either back up families—and changing families particularly—or they can split families wide apart. We ignore some of the obvious things that might be going to happen in the day-care movement.

It is in the nature of the adult to be competitive for small children. Anyone who cares about a small child automatically and unconsciously tries to shove out everybody else who might want to help with that small child. We do it as physicians with parents. We shove them out the door when we're going to give the child a shot or when we are going to take him into the hospital. Nurses push the mothers out of the room because they want to do such a good job by the child. This is in the unconscious nature of a caring adult.

A good day-care person will automatically, unconsciously, push parents out of the way; we have seen this in every day-care center that we have looked at. If we are going to change it, we must think deeply about how to focus day care on the parent-child interaction rather than on caring for the child alone. How do we do this?

We've recently been looking at day-care centers to see how pediatricians can function in them. We've been trying to find some way for pediatricians to get involved in and evaluate such centers. A most important factor in evaluating a day-care center is whether the center is able to respond to the rhythms in the child's day (states of consciousness, alertness, sleepiness, etc.). The patterns of these states are the critical issues for infants. By the age of 4 months, we found, a baby in eight-hour day care could cycle at a very low level to awake, sleep, awake, sleep—never getting very intensely involved, never getting very deep into sleep—for eight hours until the parent came at the end of the day. Then the infant would fall apart dramatically, beginning to cry, to thrash about in intense protest—what Anna Freud calls "disintegration of the ego." The child starts fussing, complaining, and turning away from the mother as soon as she arrives at night to get him. These are *the most important messages he's communicated all day*. He has saved them up for this important person. After he disintegrates, he wakes up to become his most charming, most communicative self. Inevitably, when the disintegration happens to the parent who is picking the child up, the day-care person says, "Well, he hasn't cried for *me* all day." This becomes a real putdown, because the parent blames herself for the baby's protest. She feels that it is her fault that she has had to leave the baby and that the baby is angry with her and blaming her, as she blames herself.[1] I think that no matter how necessary it is to give the baby up to day care—for whatever reasons—there is an anguish and an inevitable grief that all mothers and fathers face who have to share a small child or an infant with another adult. I call this a *grief reaction*. Part of grieving is to feel inadequate, helpless, and hopeless and to feel as if you might have already damaged the child. A common defense mechanism for these guilty and inadequate feelings is to pull farther away rather than do the child more damage. As soon as you

---

[1]See Dr. Hoover's discussion of Chapter 1, by David Brion Davis, for another perspective on this reaction of the parent.

explain this sequence of events to a day-care staff and you help them see that this sequence may cause young parents to dilute their attachment to their infant, the day-care staff change. They begin to nurture the young parents as well as the infant. They can see their own roles, as center staff, as directed toward strengthening rather than diluting the parent-child relationship and, indirectly, as strengthening the entire family. But unless we look at day care with this kind of approach, day care can become a very serious infringement on just what we are all after: deep attachments and a feeling of responsibility for small children and for the importance of the family.

In summary, I think it important that all our efforts that are directed at changing society and supporting families be looked at with just the kind of positive model for reinforcing family strengths that Dr. Sussman presented in his paper. However, I would like to add to his suggestion that we have a family-oriented national institute. I suggest that we label it or underline it by calling it a Children's Rights Institute.

7

Practical Considerations

Norman A. Scotch

I would like to shift ground a bit and follow up on some of Dr. Sussman's remarks, particularly in the sense of turning from our clinical interests to a social policy orientation. I think that we have had ample evidence about the number, the pervasiveness, and the nature of the problems associated with the family today. The question before us, then, is "As social policy people and clinicians, what are we to do about families?"

The larger perspective provided earlier by Dr. Davis sets the family in an historical context, pointing out that changes in family form and function are a part of the history of mankind. Some of the problems have been with us for centuries, in varying degrees of severity. I think that we are in a period when family problems are exacerbated, but that does not mean we should rush to tamper with the structure and functions of the family. Some of the family types have been described by Dr. Sussman as new phenomena, but many types have been with us for some time. The commune,

Norman A. Scotch • Boston University School of Medicine, Boston, Massachusetts.

for example, ranges back into history for thousands of years in a variety of cultures. It is important for us, as policy planners, both to clarify what problems are unique to today's family and to see those issues in a broader perspective. In other words, we have to have as clear a view of what is happening as is possible.

Our discussions have tended to focus on the family itself and the inner dynamics and problems that families generate. We have not talked about the family in a larger perspective except to talk about some of the forces that act upon it to create problems. I want to come back to that.

Return with me now to those golden days of Sociology 101, and ask yourselves again, as you did in those days, "What are the functions of the family, not only for its members (we've already talked about that—warmth, love, socialization, sex, reproduction, etc.), but what is the family supposed to provide for society?" When we talk about society, we're not simply talking about bureaucrats working in some agency. Society is you and me and families in our neighborhood, people we care about. What are families supposed to do? Are they supposed to provide the placements for people in the work order? Are they supposed to maintain a certain amount of stability in society? One of the things that is happening today is that we have changed our focus from the roles and obligations of the family to the needs and the problems *within* the family.

I think it's very important to take the family out of isolation and think about the obligations families have to society. When we practice our medicine and our other helping professions, we also have to remind the members of families that they have obligations as well as expectations and rights.

Now, with regard to the solution to some of these problems, much of the discussion we've had is very useful. Some of the suggestions that have been made are very, very important. But I think it is even more important for us to return to some of the points made by Dr. Kohn and Dr. McQueen. They talked about the forces acting upon the family. If we are concerned about the family, then we should be concerned with the larger social processes that are operating today.

We live in a period of great transition. Many sociologists would call this a time of anomie. And the notion of dealing with anomie and the family with Band-Aids is only one way to handle the problem. Another way is to reconceptualize the family as Dr. Sussman did. But an even better way is to look at the nature of society itself, to look at this anomic period of time we are in and to raise questions about what social processes are operating to cause the kinds of problems that we ultimately see in the clinic.

Dr. McQueen talked about the black family and about what happens to black families when larger social forces are acting upon them, for example, economic forces—economic upturns and downturns and unemployment. These are issues that deserve our attention. When we start to solve these problems of social transition—these anomic problems—then not only will we be helping the American family, but we will be turning our attention to a number of social problems—alcoholism, drug addiction, and the like. I think that it is our task not simply to focus on the family *per se* but to focus on the larger social issues as well.

Index